A GAME OF
HARE &
HOUNDS

Cover art: *Guilford Courthouse* **by H. Charles McBarron.**
Courtesy of U.S. Army Museum Enterprise art collection

An Operational-level Command Study of the Guilford Courthouse Campaign, 18 January–15 March 1781

HAROLD ALLEN SKINNER JR.

Quantico, Virginia
2021

LIBRARY OF CONGRESS CATALOGING-IN-PUBLICATION DATA

Names: Skinner, Harold, Jr., 1966- author. | Marine Corps University (U.S.). Press, issuing body.

Title: A game of hare & hounds : an operational-level command study of the Guilford Courthouse Campaign, 18 January–15 March 1781 / Harold Allen Skinner Jr.

Other titles: Operational-level command study of the Guilford Courthouse Campaign, 18 January-15 March 1781

Description: First edition. | Quantico, Virginia : Marine Corps University Press, [2021] | Includes bibliographical references. | Summary: "A Game of Hare & Hounds: An Operational-level Command Study of the Guilford Courthouse Campaign, 18 January–15 March 1781 provides a systematic analysis of this key battle in the American Revolution. This guide is intended for the instructor/facilitator, the individual who guides the execution of an applied history training event"— Provided by publisher.

Identifiers: LCCN 2020057234 | ISBN 9781732003163 (paperback)

Subjects: LCSH: Guilford Courthouse, Battle of, N.C., 1781. | Operational art (Military science) | Soldiers—United States—History—18th century. | Soldiers—Great Britain—History—18th century. | Staff rides—North Carolina—Handbooks, manuals, etc. | Southern States—History—Revolution, 1775–1783—Campaigns.

Classification: LCC E241.G9 S56 2021 | DDC 973.337—dc23 | SUDOC D 214.513:G 94

LC record available at https://lccn.loc.gov/2020057234

DISCLAIMER

The production of this book and other MCUP products is graciously supported by the Marine Corps University Foundation.

Published by
Marine Corps University Press
2044 Broadway Street
Quantico, VA 22134
First edition, 2021
ISBN 978-1-7320031-6-3

CONTENTS

MAPS AND FIGURES

FOREWORD

If we do not attack that province [North Carolina], *we must give up both South Carolina and Georgia.*
~Lieutenant General Lord Charles Cornwallis, 1780[1]

We fight, get beat, rise and fight again.
~Major General Nathanael Greene, 1781[2]

I will not say much in praise of the militia of the Southern colonies, but the list of British officers and soldiers killed and wounded by them . . . proves but too fatally that they are not wholly contemptible.
~Lieutenant General Lord Charles Cornwallis, 1782[3]

Those considering a staff ride to the American Revolutionary War battlefield at Guilford Courthouse, North Carolina, to include the key decisions shaping the overall contest in the South, have as their point of departure an indispensable resource in Harold Allen Skinner Jr.'s *A Game of Hare & Hounds: An Operational-level Command Study of the Guilford Courthouse Campaign, 18 January–15 March 1781.*

Skinner's work is solid in its research, presented in a clear and accessible style, and balanced in its treatment of issues ranging from command at the strategic and operational levels to the organization, weapons, and tactics of the two contending armies: American and British. Accepting that the term *staff ride* represents a process of going to the actual location where actions took place and critical decisions were made, all under the stress of combat and with incomplete or plain wrong information, it follows that to conduct an effective staff ride demands careful planning and preparation beforehand. To that end, this staff ride handbook is organized into clearly laid-out, highly functional components. These are part one, The Opposing Forces; part two, Southern Campaign Overview; and part three, Field Study Phase. This third phase bears special mention because it serves as a blueprint to set up the staff ride via a series of some 14 "stands" or stops, each oriented to a particular piece of terrain or vital campaign or battle stage. Parts four, Integration, and five, Support, provide a clear, cogent, and eminently workable plan for a high-value assessment of Guilford Courthouse.

If a campaign is reckoned a series of operations aimed at some specific outcome and conducted in a given expanse of space and time, then Skinner comes rightly to the vital architecture of this one: the major decisions of the two opposing commanders—Lieutenant General Lord Charles Cornwallis for the British and Major General Nathanael Greene for the Americans—and the interactions in combat of their contending armies. His approach brings us straight to the central purpose and greatest value of the staff ride, one not to be gained in any other way. Rather than being merely a review or catalog of decisions, at each of the aforementioned (in part three) stands, the author asks staff ride participants the classical questions: 1) Can you discern, at this particular stand or stage of the contest, that a decision was made (or if no decision was made, that in itself—the "no decision" —becoming functionally the decision)? 2) What exactly was the decision that was made? 3) Who made it? 4) Was it reasonable based on the information available at that time (as opposed perhaps to what is known today, after the fact)? 5) To whom was it communicated? 6) What effect did the decision have on the outcome of the event?

---

[1] Earl Cornwallis to Sir Henry Clinton, Charlestown, 6 August 1780, in Charles Derek Ross, ed., *Correspondence of Charles, First Marquis Cornwallis*, vol. 1 (London: John Murray, 1859), 54.

[2] Russell F. Weigley, *The Partisan War: The South Carolina Campaign of 1780–1782* (Columbia: University of South Carolina Press, 1970), 52.

[3] Earl Cornwallis to Sir Henry Clinton, Williamsburg, 30 June 1781, in Ross, *Correspondence of Charles, First Marquis Cornwallis*, 102.

In all this, Skinner sees the discernable beginning and ending points of the campaign as 18 January and 15 March 1781. That is, 18 January marks the aftermath of the British defeat at the Cowpens battle in South Carolina and Cornwallis's decision, in near fury, to mount a hell-for-leather in-all-weather pursuit of Greene's army as it sought to escape by falling back across North Carolina. In this so-called Race to the Dan, Cornwallis failed to trap and destroy Greene, although he destroyed wagons and precious food and rations in an effort to lighten the load and so speed up his army's rate of march. The Americans were instead able to escape to safety on the far side of the Dan River—from which point they might return to fight on some future day of their own choosing.

That is precisely what happened. The second date, 15 March, marks what proved to be the end-point of the campaign: the concluding, three-hour battle at Guilford Courthouse, where Cornwallis claimed victory—but lost one-quarter of his redcoats to death or injury. When this news crossed the Atlantic to Britain, it spurred no end of rebuke from opponents of the war against the Americans. Charles James Fox, for example, leader of the Whig opposition to King George III's Tory ministry, argued that another such "victory" might leave the British in America with no army at all. All the fruit-less marching back and forth had failed to conquer Greene, and Guilford Courthouse—far from the victorious battle Cornwallis claimed—was instead something "very nearly allied to decisive defeat."[4]

But the real situation for the British was now worse or about to prove even more so than the sarcasm could represent. The campaign and battle marked a turning point in the contest for the South. Following the loss of their only army in the South at Charleston in May 1780 (arguably Britain's single biggest victory of the war), and the routing of a second army at Camden three months later, American fortunes were now changing. After nearly a year of these severe reversals, they were swinging over to the initiative and a series of successes. In the run-up to Guilford Courthouse, for example, South Carolina militia and Continental dragoons under Lieutenant Colonel Henry "Light-Horse Harry" Lee, Robert E. Lee's father, managed to lure into ambush a sizable force of North Carolina Loyalists trying to join Cornwallis. The near annihilation of that force continued a process started at Kings Mountain the previous October, with its loss of a substantial number of the Loyalists available to the British in South Carolina. The message was clear: any Loyalists taking the field on the side of the redcoats would be joining a losing side, and they would pay the price. No further sizable body of Loyalists came forward.

And at this point the Americans became—unlike the 1780 pitched battles whereby both sides planned the encounter on a prearranged battleground—winners of a conventional pitched battle in which they were losers only in Cornwallis's dispatches. His push into North Carolina and heavy losses in battle broke the back of British efforts to bring the two Carolinas back under crown authority —a goal of London's policy from the start.

It was at this juncture that Cornwallis chose to turn the vector of British effort in the South away from his assigned task of protecting Charleston and the gains made since the previous May toward a new direction: north to Virginia. To be sure, British amphibious raids already sought to cut American lines of communications in Governor Thomas Jefferson's state, and certainly Cornwallis's initial foray netted successes, including destruction of much of the tobacco crop and capture of some number of the state's legislature. But when Cornwallis turned north, Greene turned south—and to a campaign that would roll up the British and Loyalist positions in the Carolinas and Georgia one by one and eventually regain the great prize: Charleston.

But the British were by now on the road to defeat at Yorktown, Virginia, the result of a strategic

---

[4] See in particular Cornwallis's comments dated 12 June 1781, "State of the War in America," in *Parliamentary Register or History of the Proceedings and Debates of the Second Session of the House of Commons, Fifteenth Parliament of Great Britain* (London: J. Debrett, 1782), 378. See also John W. Gordon, "Guilford Courthouse," in *South Carolina and the American Revolution: A Battlefield History* (Columbia: University of South Carolina Press, 2003), 144–48; and John Ferling, *Almost a Miracle: The American Victory in the War of Independence* (New York: Oxford University Press, 2007), 494–500.

toxic cocktail set before them by French seapower and the French and American armies acting together, as well as strategic and command discord in their own ranks. For the Franco-American allies, French admiral François-Joseph-Paul, comte de Grasse, was key, bringing his fleet from Europe via the Caribbean and up to the Virginia coast, so preventing the Royal Navy from being able to break through and rescue Cornwallis's army in its fortifications. Lieutenant General George Washington, commander in chief of the Continental Army, and Lieutenant General Jean-Baptiste-Donatien de Vimeur, comte de Rochambeau, feinted an attack on New York and then moved with dispatch south to lay siege to Cornwallis. French siege artillery pounded Cornwallis's positions. All of this occurred before British help could arrive. Cornwallis's Virginia invasion was a disaster. At Yorktown, on 19 October 1781, Cornwallis surrendered, amounting to the loss of a second British army in the field, this one coming very nearly four years to the day after General John Burgoyne's surrender at Saratoga, New York, on 17 October 1777 and effectively closing out major fighting in North America.

Staff ride participants will want to consider carefully the British command relationships at play, with overall policy and strategy for the American war being set at the ministerial level in London by Lord George Germain and overseen in America by the commander in chief in North America, Lieutenant General Sir Henry Clinton. Clinton and Cornwallis both had proven combat records, and both had seen considerable service in America.

Clinton is alternatively argued to be the most insightful British general to fight the Americans or one of the least energetic; Cornwallis is argued to be other than insightful or by far the most aggressive and energetic of the British generals to fight the Americans.[5] It is worth remembering that the war in America was not to be Cornwallis's last and that he would go on to defeat the French-backed United Irishmen's revolt in 1798, a significant threat to Britain in its home islands, and that he would die on active service in India, but only after he had greatly advanced Britain's fortunes in the subcontinent. And the contest in the South was also a competition between two British strategic approaches. One, Clinton's, argued that it was a revolutionary war that demanded the perhaps more-revolutionary (by eighteenth-century standards, at least) approach of holding enclaves and securing them by creating a Loyalist militia. Some of the Loyalists could also be formed into provincial units and brigaded with the redcoats, thus providing additional maneuver forces.[6]

Cornwallis, on the other hand, judged that the only target that mattered was the Americans' fielded forces and that these must be defeated by the most direct and decisive means available. He saw occupying Virginia as essential to this process, as it would force the Americans to come out and meet the British in decisive battle or face having the southern colonies cut off from the rest of "the States."[7]

The process of analyzing how and why all this played out as it did is the vital goal of any staff ride. Focused on a critical eight-week period in the sixth year of the American Revolution, Skinner's *A Game of Hare & Hounds: An Operational-level Command Study of the Guilford Courthouse Campaign, 18 January–15 March 1781*, makes such analysis a rewarding task.

John W. Gordon, PhD

Professor of National Security Affairs, editor in chief of *The Breckinridge Papers*

Marine Corps University Command and Staff College

---

[5] See, for example, William B. Wilcox, "Sir Henry Clinton: Paralysis of Command," and Hugh F. Rankin, "Charles Lord Cornwallis: Study in Frustration," both in *George Washington's Generals and Opponents: Their Exploits and Leadership*, ed. George Athan Billias, vol. 2 (New York: Da Capo Press, 1994), 73–102, 193–232.

[6] "British Strategy for Pacifying the Southern Colonies," in John W. Shy, *A People Numerous and Armed: Reflections on the Military Struggle for American Independence* (New York: Oxford University Press, 1976), 195–212; Piers Mackesy, "What the British Learned," in *Arms and Independence: The Military Character of the American Revolution*, ed. Ronald Hoffman and Peter J. Albert (Charlottesville: University of Virginia Press, 1984), 183–86; and Piers Mackesy, *The War for America: 1775–1783* (Lincoln: University of Nebraska Press, 1993), 251–53.

[7] But as Barbara Tuchman put it, Cornwallis thus "set out to consolidate his front by abandoning South Carolina for Virginia." Barbara Tuchman, *The March of Folly: From Troy to Vietnam* (New York: Alfred Knopf, 1984), 226.

ACKNOWLEDGMENTS

No work is done in a vacuum and I would be remiss if I did not recognize those who helped me along the path to publication. First, the staff of the Marine Corps University Press (MCUP). Director Angela Anderson and Managing Editor Stephani Miller have earned my deep respect and gratitude for turning a messy manuscript into a superbly organized and polished final product. Secondly, Marine Corps University Professor John W. Gordon provided not only a superb foreword, but his scrutiny of the historical details and interpretations improved the accuracy of the manuscript. I would also be remiss if I did not thank the Marine Corps University Foundation for their support. Mr. Charles D. Collins Jr. of the U.S. Army's Army University Press Combat Studies Institute provided a good portion of the background material for chapter 1, as well as the excellent maps used to support the text. Second, the collegial partnership with Dr. John Boyd, historian emeritus for the U.S. Army's Chaplain Regiment, and his wife, Dr. Rachel Lea Heide, defense scientist for the National Defence Headquarters of Canada, served to develop my writing and presentation skills far beyond what I could have done on my own. Honorable mentions for their help in improving my skills as an Army historian: Mr. Henry Howe of the Fort Jackson Basic Combat Training Museum; U.S. Army Reserve colonel Stephen Harlan of the 99th Readiness Division; and Colonel Craig Mix and Lieutenant Colonel Peter G. Knight, PhD (Ret), of the Army's Center of Military History. The staff at Guilford Courthouse National Military Park deserve praise for their excellent stewardship of the battlefield and assistance in answering questions about the battlefield. Particular thanks to National Park Service interpretation ranger Jason Baum and curator of the 82d Airborne Division War Museum Christopher M. Ruff, both of whom filled in many details about the armies that fought at Guilford Courthouse. Certainly, many thanks to my civilian and military leaders who indulged my passion for writing staff ride products in the course of my other historian duties. This book project was started while working at the 81st Readiness Division, U.S. Army Reserve, and many of the first edits were completed while I was deployed as the theater historian for Combined Joint Task Force-Operation Inherent Resolve. The final edits were completed while in my current position as the command historian for the U.S. Army's Soldier Support Institute at Fort Jackson, South Carolina, under the leadership of commander Colonel Stephen K. Aiton and Chief of Staff Mr. Troy Clay. I also want to thank Colonel Greg Worley and Majors JD Oblon and Justin Priestman, of the U.S. Army's Finance and Comptroller School, for the opportunity to test my manuscript during a staff ride to Guilford Courthouse with the students of the school's Captain Career Course 001-21. Deep gratitude and appreciation is owed to two of my former Army commanders, Major General David C. Wood and Brigadier General Brian R. Copes. Their willingness to let me work as a unit historian, both at home and downrange, set me up for an enjoyable and rewarding post-retirement career as a civilian historian. Last, but not least, thanks are due to my wife Jackie and our children, who patiently indulged my interest in all things history-related in countless field trips to museums and battlefields. Naturally, all mistakes or errors of fact and interpretation are mine. Colossians 3:23.

INTRODUCTION

During the afternoon of 15 March 1781, two armies—one comprising British and German regulars, the second a collection of American Continental Army regulars, state troops, and militia—fought the pivotal Battle of Guilford Courthouse. On the surface, the battle was a remarkable tactical victory for the British. When viewed within the context of the British strategy in the Southern Campaign, the outcome of the Guilford Courthouse battle was a victory devoid of strategic benefit. Lieutenant General Lord Charles Cornwallis's army was rendered incapable of consolidating any gains, and without the security presence of British regulars, Loyal Americans (also known as Loyalists or Tories) were not willing to risk their lives and property in service to the crown. Cornwallis subsequently abandoned North Carolina, thereby ending the crown's last realistic attempt to gain and retain American territory sufficient to serve as leverage when the time came for peace negotiations. Conversely, Major General Nathanael Greene displayed an impressive grasp of linking operational design to strategy, which, as a consequence, gave the Americans ultimate control of the Carolinas—despite his dismal tactical record while in command of the Southern Department, the Continental Army's organizing unit for regiments of Virginia, the Carolinas, and Georgia. The study and analysis of the Guilford Courthouse battle can and should reveal insights at the three levels of war: strategic, operational, and tactical.

The importance of the Guilford Courthouse battle is best understood by briefly reviewing the state of the British Southern Campaign in June 1780. Orchestrated by Lieutenant General Sir Henry Clinton, commander in chief of the British forces in America, expeditions during the winter of 1779–80 had won a series of impressive victories, taking Savannah, Augusta, and much of the Savannah River valley between South Carolina and Georgia. Clinton scored one of the largest British victories in the entire war when a joint army-navy expedition captured the major port city of Charleston, along with its Continental garrison of approximately 3,300 troops, in May 1780. Captured Patriot militia were paroled and sent home, and at first, pacification efforts seemed to go well, with British officers and cadre hard at work organizing Loyalist militias to secure vital logistics points. From that apparent point of complete dominance, British fortunes in the South rapidly deteriorated. British arrogance and ineptitude in the treatment of rebel sympathizers and undecided "fence-sitters" drove many paroled Patriot fighters back into active resistance. Small Patriot partisan bands, mostly under the direction of Francis Marion (the "Swamp Fox") and Thomas Sumter (the "Gamecock") attacked isolated British outposts and ambushed Loyalist detachments. Most skirmishes were won by the Patriots, and by late summer the rebels had gained a marked moral advantage over their Tory enemies.

Concerned with the deteriorating security situation, Cornwallis tried a risky expedient, sending a corps of the best Loyalist militia regiments, led by Scottish regular Major Patrick Ferguson, into the rebel sanctuary areas along the Blue Ridge Mountains. Aroused by the threat to their settlements, Patriot militia regiments formed a punitive expedition that annihilated Ferguson's militia corps at Kings Mountain on 15 October 1780.

The stunning Patriot victory at Kings Mountain not only reversed the string of tactical successes by the British, but "spirit[ed] up the people" opposing the British occupation.[1] Bereft of his best militia regiments, Cornwallis suspended his offensive drive into North Carolina to return his regular units to pacification duties back in South Carolina. Greater trouble for Cornwallis came when Major

---

[1] Gen Nathanael Greene letter to Col Daniel Morgan, "The Detached Command Formed Which Won Cowpens," 16 December 1780, in Theodorus B. Myers, *Cowpens Papers, Being Correspondence of General Morgan and the Prominent Actors* (Charleston, SC: News and Courier Book Press, 1881), 9, hereafter Greene to Morgan, 16 December 1780.

General Greene took command of the threadbare Southern Department in late 1780. Selected by General George Washington to replace the incompetent Major General Horatio Gates after his defeat at the Battle of Camden, Greene proved an inspired choice for command, as he was not only a competent logistician, but a gifted strategist. After gaining situational awareness and understanding of the southern theater, Greene implemented a Fabian strategy, designed to minimize his weaknesses while neutralizing Cornwallis's greater combat strength.[2] First, Greene adopted a delaying strategy to preserve his precious Continental units, attacking only isolated British detachments and avoiding decisive battle with Cornwallis's superb regular units. Although Greene struggled at times in his relationships with local Patriot militia leaders, particularly with Thomas Sumter, Greene was able to employ Whig partisans to harass British forward bases and lines of communication. By adopting an indirect operational approach, Greene expected to weaken Cornwallis enough to prevent British consolidation of political gains in South Carolina; ideally, enough to force Cornwallis to abandon the Carolinas altogether.

Greene certainly had his own challenges, with the lack of supplies around Charlotte, North Carolina, and the general turbulence in his militia units. In another stroke of brilliance, Greene turned his logistics problem upside down and, in the process, utterly confounded Cornwallis. Greene split his army, ordering the light infantry and dragoons under the command of Brigadier General Daniel Morgan across the Broad River in northwestern South Carolina. Meanwhile, Greene marched the remaining Continentals and logistics trains to Cheraw Hill in far northeastern South Carolina. Ordering Morgan so far west accomplished several purposes. First, Morgan could draw his own provisions from the region, reducing the logistics strain on the Continental Army, while simultaneously denying the same resources to British units. Second, the presence of armed Continentals served to "spirit up" the local Patriots, and suppress potential and actual Tory sympathizers.[3] Most importantly, Greene's dispositions denied Cornwallis the freedom of maneuver, as a British advance toward either army would expose vulnerabilities for the other American army to attack. Goaded into action by Morgan's incursion, Cornwallis attempted to trap Morgan's flying army (Continental infantry and dragoons supported by light infantry militia; see Selected Glossary for more information), sending Lieutenant Colonel Banastre Tarleton's British Legion to the west in pursuit. Tarleton's overaggressive advance ended in a total defeat at the Cowpens on 15 January 1781.[4] Disrupted and encircled by Morgan's innovative three-line area defense, Tarleton lost around 80 percent of his 1,100 dragoons and infantry to death or capture.

The back-to-back defeats at Kings Mountain and the Cowpens provoked Cornwallis into abandoning his pacification mission in an all-out pursuit of Greene's consolidated army in what is now known as the Race to the Dan. By early February 1781, Cornwallis's command was hungry and tired, stranded in the destitute Dan River valley along the North Carolina-Virginia border with no effective Loyalist support. Cornwallis stubbornly refused to abandon his pursuit, and Greene skillfully kept the British Army at arm's length across the Dan River, while his army refitted and received militia and state line reinforcements from Virginia and North Carolina. Although he was loath to risk his Continentals in battle, Greene realized that he had to fight Cornwallis to keep his regulars from returning to pacification tasks. Once Greene felt he had received sufficient supplies and reinforcements, he crossed the American force back into North Carolina in late February 1781. Even in close proximity to the enemy, Greene carefully moved his force to remain just out of reach of Cornwallis's army until the British supply situation became critical. Finally, Greene camped his force of militia and Conti-

---

[2] Greene is even credited with convincing Gen Washington to adopt such a strategy of wearing down an opponent through a war of attrition and indirection. See Gregory J. Dehler, "Fabian Strategy," Washington Library, George Washington's Mount Vernon (website), accessed on 13 November 2018.
[3] Greene to Morgan, 16 December 1780.
[4] Richard L. Morgan, *General Daniel Morgan: Reconsidered Hero* (Morganton, NC: Burke County Historical Society, 2001), 22.

nentals at Guilford Courthouse, allowing Cornwallis to move to contact on 15 March 1781. The ensuing battle resulted in a narrow tactical win for the British field army, which retained possession of the ground and the abandoned Continental artillery pieces, while Greene successfully extricated his Continental regiments without severe manpower losses.

In the end, Cornwallis won the battle but, in the process, had driven his command to and beyond its offensive culmination point, as its regiments were badly hurt and could no longer effectively provision themselves in the region. Consequently, Cornwallis withdrew his command to the friendly seaport of Wilmington, North Carolina. Greene swiftly took advantage of Cornwallis's withdrawal by marching back into South Carolina to reduce the remaining British bases. Although the Southern Army never won a single major battlefield victory during the remainder of the war, Greene's indirect approach in leading the campaign ultimately drove the British from interior South Carolina. By the end of the war, the British were penned in their coastal enclaves around Charleston and Savannah, powerless to protect Loyalist supporters. Even in the dark days immediately after the Cowpens, and facing pursuit by an implacable enemy, Greene accurately envisioned the outcome of the Southern Campaign: "It is necessary we should take every possible precaution to guard against a misfortune. But I am not without hopes of ruining Lord Cornwallis, if he persists in his mad scheme of pushing through the Country. . . . Here is a fine field and great glory ahead."[5]

Greene was not an exceptionally good battlefield commander, especially when compared to Morgan and Cornwallis. For example, Greene neglected much of Morgan's thoughtful advice in handling militiamen by not giving them a realistic mission and clear commander's guidance before the Battle of Guilford Courthouse. Greene lacked the ability to motivate and inspire the militia by direct example; after giving a tepid speech to the North Carolina militia at the first line, Greene retired to the Continental third line before the battle. Lacking a firm command presence, the first line collapsed quickly when confronted with British bayonets, and few North Carolinians remained to support the rest of the battle. Unlike Morgan's relatively compact deployment at the Cowpens, Greene deployed his individual lines too far apart for mutual support, thus leaving his army open to defeat in detail. Greene's genius is seen in his firm focus on operational goals, weakening Cornwallis's army while avoiding the destruction of his Continental regiments. In that respect, Greene performed admirably. From the time his army crossed the Dan until the Battle of Guilford Courthouse on 15 March 1781, Greene played a daring game of hare and hounds, maintaining the initiative and forcing Cornwallis to react.[6] Greene personally selected advantageous terrain near Guilford Courthouse and ensured his force had a well-thought-out plan in the event of a withdrawal.

By comparison, General Cornwallis comes across as a hard-fighting tactical commander who, despite having a professional military education, could not grasp the operational and strategic implications resulting from his tactical decisions. This point is probably best illustrated by Cornwallis's decision to burn most of his logistics train during the pursuit of Greene's army. From a tactical standpoint, the move allowed Cornwallis's army to move faster and reduced the number of fighters needed to secure the trains. From an operational standpoint, the decision gravely impacted the ability of the British Army to sustain itself at long distance from the closest supply depots. Furthermore, Cornwallis did not study the British failures at Kings Mountain and the Cowpens to distill tactical lessons for future battles with the Americans. At Guilford Courthouse, Cornwallis used essentially the same linear infantry tactics as was used at Camden, even so much as employing his light infantry in the battle line

---

[5] Nathanael Greene: Gen Greene to BGen Isaac Huger, 30 January 1781, in *The Papers of General Nathanael Greene*, vol. 7, *26 December 1780–29 March 1781*, ed. Richard K. Showman, Dennis M. Conrad, and Roger N. Parks (Chapel Hill: University of North Carolina Press, 1994), 220. Greene's remark came after he learned of Cornwallis's burning of the baggage wagons, lightening the load of the army for the planned pursuit of Greene's army.
[6] *Hare and hounds* refers to an early 1800s outdoor game in which some of the players leave a trail and others try to follow the trail to find and catch them; also a portable game of the same name and concept played with a board and pieces.

instead of performing their traditional battlefield screening and skirmishing tasks. Furthermore, the British Legion, crippled by the loss of its light infantry and the wastage of its best horses, was largely unable to penetrate the American dragoon screens to gather actionable intelligence about American strengths and dispositions. As a consequence of such leadership failures, British and Hessian units repeatedly blundered into American kill zones, and suffered heavy casualties at Guilford Courthouse.

Although the battle took place more than 236 years ago, study of the Guilford Courthouse battle set within the context of the Southern Campaign will reveal many valuable insights into the operational and strategic levels of war. Many lessons are to be found at the tactical level as well. Of particular interest to junior leaders are the vignettes that highlight the exercise of leadership in combat, and how the participants faced their moment of destiny in battle.

PLANNING AND ORGANIZATION

A Game of Hare & Hounds: An Operational-level Command Study of the Guilford Courthouse Campaign, 18 January–15 March 1781 provides a systematic analysis of this key battle in the American Revolution. Part one provides a basic description of the American and British armies: weapons, tactics, logistics, communications, engineering, intelligence, and medical support. Part two offers a campaign-level overview followed by a detailed description of the battle. Part three recommends an itinerary of sites (or stands) to visit on the battlefield. Each stand is organized with directions, orientation to the battle site, detailed description of the action that occurred at the location, and vignettes drawn from participants in the battle, and each concludes with suggested questions for analysis. Part four provides an outline for the integration phase, in which students synthesize their classroom and field phase learning, ideally to glean relevant lessons for use in their military profession. Part five of this guide is written for the operations and logistics staff planners who will handle all of the administrative, training, and logistics coordination inherent with any good training event. Note, the author has included many direct quotes from primary source material throughout; original word and name spelling and capitalization variations have been retained without individual notations except where absolutely necessary for clarity. appendix A gives thumbnail biographical sketches of the major participants, appendix B provides a detailed order of battle, and appendix C gives a chronological account of the campaign. Finally, a selected bibliography is included with recommended sources for additional study before the terrain walk. The bulk of this guide is intended for the instructor/ facilitator, the individual who guides the execution of an applied history training event. To do so, the facilitator must become thoroughly familiar with the material, which is best done in conjunction with a personal reconnaissance to walk the battlefield and understand the relationship between the physical terrain features present at the time of the battle, and the historical events as they unfolded on the ground.[7]

Before conducting the classroom study and actual battlefield terrain walk, the instructor should provide a recommended list of read-ahead materials so students can gain a basic knowledge of the campaign and battle. The selected bibliography in the back of this guide is an excellent starting point, as it includes many digital documents easily downloaded at no cost to the student. Fortunately, a fair amount of primary source material, in the form of eyewitness accounts and reports, is available, which greatly helps in understanding the human dimensions of the battle. Besides this handbook, the author recommends the detailed National Park Service (NPS) study, *The Battle of Guilford Court-*

---

[7] Curtis S. King, "The Staff Ride" (PowerPoint presentation, Military History Instructor Course, U.S. Army Combat Studies Institute, Fort Leavenworth, KS, 2013), 1.

house written by Charles E. Hatch Jr., which is available at no cost from the NPS history publications repository.[8]

Individual study by the student is followed by an instructor-facilitated classroom study that imparts basic historical and operational knowledge. To best increase buy-in and participation, the instructor should use a seminar format that assigns students to present a short functional briefing that describes a particular facet of the battle, such as major battlefield personality, warfighting function, branch or functional area, or major events before or during the battle.

The terrain walk portion of the staff ride should take no more than eight hours, as the battlefield (less the visitor center video and terrain map) is easily traversed on foot. The battlefield is relatively compact and encompassed by a paved multiuse road 2.25 miles (3.62 kilometers [km]) long. By foot, the walking distance from the American first line to the final battle line is slightly more than 1 km. The battlefield is fairly well preserved in its original condition, with some significant exceptions. First, several monuments were installed after the battle that have no connection to the events of the battle, and if not accounted for by the staff ride facilitator, these can serve as a distraction during the terrain walk. The muddy fields traversed by Cornwallis's troops in their first line attack are today covered in grass. The area west of the Joseph Hoskins farmstead, where Cornwallis initially deployed his troops, and around New Garden Meeting House, which saw the opening skirmish between Tarleton and Lieutenant Colonel Henry "Light-Horse Harry" Lee III, are now part of the urban sprawl of Greensboro, North Carolina. Most significantly, the open terrain to the west of the American third line, the portion that stretches from NPS marker 7 to 5, is flanked by a large marker that incorrectly identifies the area as the American third line. Later NPS scholarship places the position of the American third line about 400 meters uphill to the east-northeast, closer to the old courthouse.[9] At the time of the final battle at the third line, much of the vale near the creek had been a cleared farm plot gone fallow, but today the intervening area is thickly wooded, which can cause some difficulties for participants trying to visualize the final engagements on the third line.

The sequence of stands in the guide is ordered to facilitate a logical flow with minimal backtracking; but the facilitator can easily add, modify, or delete stands as needed to best support the training objectives and time available. Each stand is written incorporating the U.S. Army Combat Studies Institute's (CSI) staff ride logic structure: orient, describe, and analyze (ODA).[10] First, provide orientation to the terrain and physical conditions (time, weather, and lighting) present *at the time of the battle*. Next, the instructor should lead a description of a particular action or aspect of the battle, when possible using the included vignettes from participants of the battle to illuminate the face of battle. Particularly useful here are role players describing their decisions and actions during the battle. Finally, an analysis portion should be included to provide time to allow students a chance to share insights derived from critical analysis, instructor-led synthesis of concepts and ideas, and a linkage to possible modern applications. At the end of the terrain walk comes the critical point of the entire event, the integration phase, which should take place shortly afterward, permitting students to capture, synthesize, and orally articulate their observations and insights. To omit or rush this portion is to miss the entire point of the staff ride: What did I learn, and how do I apply what I learned today to improve myself and/or my profession?[11]

---

[8] Charles E. Hatch Jr., *Guilford Courthouse National Military Park: The Battle of Guilford Courthouse* (Washington, DC: U.S. Department of the Interior Office of History and Historic Architecture, 1971).

[9] Lawrence E. Babits and Joshua B. Howard, *Long, Obstinate, and Bloody: The Battle of Guilford Courthouse* (Chapel Hill: University of North Carolina Press, 2009), 238. Interpretive NPS markers have been updated to reflect the more current interpretation.

[10] King, "The Staff Ride," 10, 13.

[11] Matthew Cavanaugh, "The Historical Staff Ride, Version 2.0: Educational and Strategic Frameworks" (master's thesis, United States Military Academy, West Point, New York, 2013), 4–6. Staff ride facilitators are strongly encouraged to read Cavanaugh's paper as a means of ensuring the staff ride is conducted with sufficient historical rigor.

A Game of Hare & Hounds

PART I

THE OPPOSING FORCES

Before starting the Guilford Courthouse campaign, it is important that the reader understands the broader strategic and operational context in which the campaign took place. This first chapter will provide the reader with a concise survey of the organization, weapons, tactics, and support functions of the British and American armies, details that provide context to the historical events and leader decisions covered in the later chapters. For a more in-depth study of the subject, the reader is encouraged to consult the references in the selected bibliography, beginning with Charles E. Hatch Jr.'s study, *Guilford Courthouse National Military Park: The Battle of Guilford Courthouse*, which is available free of charge via the National Park Service's digital reference library.[1]

UNIT ORGANIZATION

Prewar Colonial America

Before the American Revolution, the standing militia consisted of all able-bodied White males, aged 16–60, who were required to periodically assemble, or muster, for training with a musket, ammunition, and basic supplies. In times of threat from hostile intruders, militia were liable for service up to 90 days, but in practice the colonial governor would discharge the militia as soon as the crisis was passed. The militia was deemed sufficient to provide both internal and external security, so few British regular troops were stationed in America prior to the French and Indian War of 1754–63. During that conflict, a small number of Americans were enlisted as volunteer (Provincial) troops, with the majority of support given to the British Army from regular militia who were employed as scouts, guides, and skirmishers. Although a victory for the British, the war produced a wedge between American militia and British regulars: the Americans felt the British were arrogant and condescending, while Major General Edward Braddock's defeat at the Battle of the Monongahela on 9 July 1755 exposed rampant ineptness within the British establishment. In turn, British regulars disdained the lack of discipline in the American militia ranks. Disagreements over the sharing of wartime costs, particularly the quartering and feeding of the regulars, was a contributing cause of the American Revolution in April 1775.[2]

Birth of the Continental Army

After the Middlesex County battles at Lexington and Concord in April 1775, New England militia units converged on Boston to reinforce the Massachusetts regiments besieging the town. Each state had its own commander in chief and problems with unity of command soon became apparent. So, on 14 July 1775, the Second Continental Congress created a new national army by mustering state militia regiments into service for a six-month period. To command the new Continental Army, Congress selected Colonel George Washington, a Virginia militia colonel with combat experience during the French and Indian War. By March 1776, General Washington and Congress had organized a Continental Army with an authorized strength of 13,000 officers and soldiers in 27 regular infantry regiments. To exercise theater-level operational command, Congress created three departments,

---

[1] Charles E. Hatch Jr., *Guilford Courthouse National Military Park: The Battle of Guilford Courthouse* (Washington, DC: U.S. Department of the Interior Office of History and Historic Architecture, 1971).
[2] Richard W. Stewart, ed., *American Military History*, vol. 1, *The United States Army and the Forging of a Nation, 1775–1917*, 2d ed. (Washington DC: U.S. Army Center of Military History, 2009), 30–32.

each commanded by a Continental major general. Significantly, the department commander lacked command authority over state militia units, unless explicitly granted by the state governor.[3]

The infantry regiment was the largest permanent tactical unit in the Continental Army. Commanded by a colonel and assisted by a lieutenant colonel and major, the regiment was authorized eight companies, each with 1 captain as commander, 4 other officers, 8 noncommissioned officers (NCOs), 2 musicians, and 76 privates, totaling 728 officers and soldiers. Enlisted soldiers were generally drawn from the lower classes of society, and their length of enlistment varied according to the fortunes of war.[4] By contrast, most officers came from the prewar gentry or mercantile class and were usually commissioned for the duration of the war. For field service, regiments were grouped into brigades or divisions commanded by a senior colonel or brigadier general, which were in turn combined into a field army commanded by a major general. Staff roles were filled by detailed regimental officers. One such example was Lieutenant Colonel Otho Holland Williams, who functioned both as an adjutant for the Southern Department and deputy commander of the 1st Maryland Regiment. Most logistics functions, quartermaster, and transportation activities were handled by skilled civilian contractors.

The high cost of recruiting, training, and sustaining both horses and dragoons meant the Continental light dragoon establishment was limited to four regiments, each authorized 280 officers and troops. In a cost-avoidance move, many states raised their own militia dragoon troops, which were cheaper to maintain than the full-time Continental troops. Dissatisfaction with the poor self-defense capabilities of the dragoons led Washington in 1780 to reorganize the dragoon regiments into legionary corps by converting two troops into light infantry. The change not only created a better-balanced light unit, but with 100 fewer horses to feed it produced a substantial cost savings for the cash-strapped Americans. The success of the legionary corps led to further reorganization to create partisan corps, each authorized four troops of dragoons and four companies of light infantry.[5]

Despite the relative lack of artillery units in the colonial militias, the Continental Army was able to field a fairly robust and effective artillery force under the leadership of Major General Henry Knox.[6] In 1776, the Continental Army organized a 12-company artillery regiment using captured British cannons and a cadre of ex-Royal artillerymen from the French and Indian Wars. The regiment served as an administrative headquarters to the artillery companies, with each authorized 6 officers, 8 NCOs, 9 bombardiers, 18 gunners (ranked as privates but paid extra as specialists), and 32 matrosses (artillery privates ranking below gunners). During campaigns, artillery companies were usually attached in direct support to Continental infantry brigades.[7]

Providing specialized logistics support to the Continental Army were artificer companies supervised by the department's Quartermaster Department.[8] In the Southern Department, Major General Greene established an intermediate logistics depot at Salisbury, North Carolina, to support operations in the Carolinas and Georgia. Salisbury functioned as a distribution point for supplies

---

[3] Steven E. Clay, "Staff Ride Guide to the Saratoga Campaign" (unpublished draft manuscript, Combat Studies Institute, U.S. Army Combined Arms Center, 2017), 3–5. The length of detail varied from 90 days for traditional militia unit up to 18 months for state line units, such as the Virginia Line units that fought at Guilford Courthouse.

[4] Robert K. Wright Jr., *The Continental Army*, Army Lineage Series (Washington, DC: U.S. Army Center of Military History, 2006), 68–69.

[5] Wright, *The Continental Army*, 161–62. A common eighteenth-century alternative spelling of *partisan* was *partizan*, which students may encounter in further readings of contemporary works. See John C. Fitzpatrick, *The Writings of George Washington from the Original Manuscript Sources, 1745–1799*, vol. 20 (Washington, DC: Library of Congress, 1776), 278. The standard modern spelling is used throughout this work.

[6] Clay, "Staff Ride Guide to the Saratoga Campaign," 6–8.

[7] Wright, *The Continental Army*, 53–54.

[8] Risch Erna, *Supplying Washington's Army* (Washington, DC: U.S. Army Center of Military History, 1981), 25, 33. *Artificers* were skilled craftsmen who provided important logistics support. *Quartermaster artificers* built and maintained camp barracks, wagons, and bateaux (flat-bottomed boats) and in the field assisted combat troops in building field fortifications. *Artillery artificers* had similar carpentry skills, used in building artillery carriages and wagons, but were trained to repair cannon and small arms.

and ordnance stores, as well as providing equipment maintenance and manufacturing support to Continental units in the region.[9]

The American Army: State Troops

Theoretically, each state adopted the Continental regimental structure for their militia regiments, but in practice the units varied in organization, method of recruitment, and length of service. As Greene was to realize, Patriot militia units came with considerable advantages, as well as significant limitations. Theoretically, militia would report for field duty with the basic arms and accoutrements, requiring little more than daily rations and an occasional resupply from Continental ordnance stores. In reality, many of these troops reported for duty missing vital equipment and even weapons, so Greene's quartermasters had to keep stocks of muskets, uniforms, and accoutrements for resupplying state troops. When around their home districts, Patriot militia provided invaluable intelligence concerning the terrain and loyalty of the population and would often fight harder in the defense of their homes and townships. A significant number of militia officers and NCOs had prior experience in frontier or conventional wars and some even had prior Continental experience. When intelligently led and employed according to their capabilities, militia units were potent force multipliers for the Continental Army—best exemplified by Brigadier General Daniel Morgan's skillful employment of militia at the Cowpens in January 1781.[10] However, Patriot militia had significant weaknesses. If well-led, a militia might stand for a time under heavy fire, but it would invariably give way to a British bayonet assault, as these troops were seldom trained or equipped for bayonet fighting. The motivation and effectiveness of the militia units tended to drop the farther they marched from their home districts. Even with quality leadership, discipline in militia units was seldom good, and a battlefield reverse would often result in widespread desertions. Finally, the predilection of militia to bring their own horses and their demands for fodder further burdened the already strained Continental quartermaster system.[11]

The Continental Southern Department, 1779–81

First organized in February 1776, the Southern Department evolved into a backwater after then-major general Sir Henry Clinton's bungled attempt to take Charleston, South Carolina, in April 1776. The uncoordinated British attempt to force Charleston Harbor failed partially due to the stubborn American resistance, but principally due to the failure of the British Army and Royal Navy to coordinate their actions. Emboldened by the British failure, southern Patriots largely suppressed Loyalist (Tory) militia organization efforts within the region. The surprise capture of Savannah, Georgia, in December 1778 alerted the Americans to the British shift in strategy toward the southern colonies. Congress scraped together reinforcements and by early 1780, department commander Major General Benjamin Lincoln had a force of 3,500 troops defending Charleston. Lincoln's force included all of the Continental infantry regiments from Georgia and the Carolinas, state regiments from Virginia, and a small contingent of regular and militia dragoons. After an extended campaign and siege, marked in large part by American ineptness, now-lieutenant general Clinton's army captured Charleston on 12 May 1780. Clinton had learned from the mistakes of 1776 and employed an indirect approach to isolate Charleston from its vulnerable landward side and used capable liaison

---

[9] Lawrence E. Babits and Joshua B. Howard, *Long, Obstinate, and Bloody: The Battle of Guilford Courthouse* (Chapel Hill: University of North Carolina Press, 2009), 25.
[10] Wright, *The Continental Army*, 166.
[11] Wright, *The Continental Army*, 67–75.

officers to prod the reluctant Royal Navy commodore to effectively support the British Army. Further disaster befell the American cause when the last remaining Continental detachment in South Carolina was destroyed at the Waxhaws on 29 May 1780 by Lieutenant Colonel Banastre Tarleton's *British Legion*.[12]

The Continental Congress responded to the Charleston debacle by reconstituting the Southern Army in North Carolina, using the reinforced Maryland-Delaware Division as a nucleus. Congress overruled General Washington's objections to appoint Major General Horatio Gates, the hero of the Saratoga campaign, as the department commander. Disregarding sound advice from his subordinates, Gates then marched his little army through the pro-Tory pine barrens of central South Carolina to attack the poorly defended British base at Camden. Warned of Gates's approach by Loyalist sympathizers, General Lord Charles Cornwallis consolidated his scattered regiments and marched to confront Gates. Advanced elements of both armies fought a brief meeting engagement on the night of 14 August 1780. The next morning, Gates badly fumbled his combat deployment, placing his militia regiments so that they were easily shattered by Cornwallis's best regiments, who in turn outflanked and crushed the outnumbered Continentals, killing 900 and capturing 1,000 prisoners along with the American artillery park and logistics trains. Assuming that American resistance was largely crushed, Cornwallis decided to resume his advance into North Carolina after the fall harvest. Cornwallis's plans were upended when a band of Patriot militia demolished his best militia forces at Kings Mountain, South Carolina, in October 1780.

Stunned by the disaster and plagued by sickness in the ranks, Cornwallis suspended his offensive and withdrew his army to winter quarters around Winnsborough, South Carolina.[13] Meanwhile, in December 1780, Major General (and Quartermaster) Nathanael Greene rode into Hillsborough, North Carolina, and quietly relieved Gates from department command. Greene reviewed his command to find only 905 Continentals and 1,500 militia—many sick and all dispirited and hungry. After performing a reconnaissance of his new department and briefly pondering his dismal circumstances, Greene moved to seize the initiative. Mindful of his weak army, Greene deliberately planned to divide it to confound Cornwallis, thus keeping the British from resuming the offensive into North Carolina. First, Greene sent the flying army of 1,200 light infantry and dragoons under command of Brigadier General Morgan west of the Catawba River to threaten the British supply base at the town of Ninety Six, South Carolina. Meanwhile, Greene marched the rest of his army into the pro-Patriot Cheraw Hill region in South Carolina, where he could safely subsist his command while threatening Cornwallis's lines of communication to the coast.

Greene's strategy of dividing his army triggered a violent reaction from Cornwallis, who in turn split his army in an attempt to trap and destroy Morgan's light corps. Instead, Morgan set up a well-planned defense at the Cowpens, and decisively defeated Lieutenant Colonel Tarleton's *British Legion* there on 17 January 1781. Afterward, Greene consolidated his army and withdrew toward North Carolina with hundreds of British prisoners in tow. During several weeks in January and February 1781, Greene orchestrated an extended delaying action (which included evacuating the Salisbury depot) that ended when the Americans reached sanctuary in southern Virginia with no appreciable losses. There, Greene refitted and reorganized his army, which consisted principally of Brigadier General Isaac Huger's Virginia Continental Brigade of 900 fighters and Colonel Otho Holland Williams's Maryland Continental Brigade of 800 soldiers. In support of the Continental infantry were 210 dragoons, around 600 light infantry and riflemen, and a Continental artillery company of four

---

[12] Babits and Howard, *Long, Obstinate, and Bloody*, 6.
[13] John W. Gordon, *South Carolina and the American Revolution: A Battlefield History* (Columbia: University of South Carolina Press, 2003), 92–95.

6-pounder cannons. Greene was still critically short of militia personnel, with Colonel Andrew Pickens's Carolina militia regiment of 200–300 troops as his best contingent. While the Continentals rested and refitted, Greene called for more militia reinforcements.[14]

The British Army: Regulars, Provincials, and Militia
Regulars

By 1780, Lord George Germain, secretary of state for the American colonies, was in overall control of British grand strategy in North America, which was divided into two defined geographic commands. Clinton functioned as commander in chief and governor-general of all British-controlled land from the south of the Saint Lawrence River down to the Gulf of Mexico.[15] As in the Continental Army, infantry regiments were the largest permanent tactical unit in the British Army and were normally paired in temporary brigades under a field army commanded by a major or lieutenant general. British regiments were commanded by a colonel assisted by a lieutenant colonel, a major, and a small specialist staff, and were nominally composed of 12 infantry companies. When a regiment deployed to North America, two companies were left behind in Britain for recruiting, leaving eight line infantry and two flank companies, one of light infantry and the other a grenadier company. Originally, grenadier companies were formed by tall and strong infantry employed in assaulting fortifications with black powder grenades. By 1776, the grenadier function was obsolete, but the name was retained as a mark of honor for the infantry company deployed on the right of the regiment, traditionally the point of decisive action in combat. Light infantry companies were organized and trained to provide skirmishers, when the regiment was not traditionally deployed to protect the left of the regiment. The grenadier and light infantry companies naturally attracted the best soldiers in the regiment, so they were often grouped into ad hoc assault battalions under division or army control. Compared to the theoretical fighting strength of 544 muskets in a Continental regiment, a British regiment organized for combat fielded only 448 muskets.[16]

The high cost of shipping and maintaining horses in North America meant the British limited their mounted troop strength in the American colonies to two dragoon regiments. By comparison, the British Army was well supported by the *Royal Artillery Regiment*, which was considered a separate and coequal military service. *Royal Artillery* battalions served as a purely administrative echelon, with the artillery company as the primary tactical echelon. Artillery companies had a standard organization that varied according to the availability and type of cannon systems. Cannons were useful for their moral effect on the battlefield but were heavy and costly to move, a factor that limited British field artillery employment in the Southern Campaign to 3- and 6-pounder cannons.[17]

The British Army clearly held many tactical advantages over the Continental Army, as it was a well-established professional military force with a long tradition of battlefield victory. Many of the officers and NCOs were combat veterans, but regardless of combat experience, all were capable of quickly molding new recruits into professional soldiers. As a consequence, the Continental Army fought at a decided disadvantage in the early years of the war, not reaching tactical parity until the Battle of Monmouth near Monmouth Court House, New Jersey, in June 1778. However, the tactical prowess of the British Army was constrained by significant strategic weaknesses, particularly logistics,

---

[14] Babits and Howard, *Long, Obstinate, and Bloody*, 220–21.
[15] David K. Wilson, *The Southern Strategy: Britain's Conquest of South Carolina and Georgia, 1775–1780* (Columbia: University of South Carolina Press, 2005), 59–60.
[16] Wright, *The Continental Army*, 47–49. The actual combat strengths of infantry regiments were generally 40–60 percent of authorization due to sickness and details.
[17] Clay, "Staff Ride Guide to the Saratoga Campaign," 19–22.

as the supply lines between England and North America were long and vulnerable to enemy interdiction. In addition, recruiting of native British volunteers was difficult, as soldiering was not considered an honorable profession and lurid tales of combat in the wilds of North America dissuaded many volunteers. Even after the depot regiments collected a sufficient number of new recruits, many were bound to die by disease or accident before even reaching their regiments in the field. As a consequence, British infantry regiments in North America were chronically undermanned throughout the war.[18]

Provincials

The practice of enlisting Americans in provincial regiments dates back to the French and Indian War and was revived as a means of offsetting shortages of regular units. Provincial regiments enjoyed the same pay and benefits as the regulars but were limited to service only for the duration of the war. In the hierarchy of the British Army, provincial officers were lower in status than regulars and were not entitled to half pay and permanent retention of rank after demobilization. Many of the provincial units to serve in the Southern Campaign were actually raised in New York and New Jersey, including one such provisional corps of volunteers organized by Major Patrick Ferguson in late 1779. After the capture of Charleston, Ferguson's provincials were used to organize and train Carolina Loyalist militias. In October 1780, Ferguson's provincial corps and some 800 of his best militia were killed or captured at the Battle of Kings Mountain. Left untouched by the disaster was the 150-person *Royal North Carolina (Provincial) Regiment*, commanded by Lieutenant Colonel John Hamilton, which remained with General Cornwallis's army throughout the Southern Campaign, serving as the guard force for the British logistics echelon.[19]

As noted earlier, the British lacked sufficient light dragoons for intelligence, patrolling, and flank security missions. Lieutenant General Clinton prodded local Loyalists to raise additional provincial units, and by July 1778, the *British Legion* was organized with 250 dragoons and 200 light infantry under the field command of Lieutenant Colonel Tarleton. For the Southern Campaign, additional infantry units and a 3-pounder cannon section were attached to the *British Legion* to create a robust combined arms regiment of 1,100 fighters. However, Tarleton's tactical ineptitude led to the loss or capture of most of his command at the Battle of Cowpens in January 1781, reducing the *British Legion* to a remnant of 272 mounted soldiers.[20]

Loyalist Militia

The British expended considerable effort during the summer of 1780 in organizing district militia regiments as part of the overall pacification strategy, with Major Ferguson and his provincial troops as trainers and cadre. In late August 1780, Cornwallis ordered Ferguson to deploy with his best militia regiments into the Blue Ridge piedmont to expand his pacification and recruiting efforts. Aroused by the Loyalist presence so close to their settlements, the Overmountain men formed an expedition that surrounded and destroyed Ferguson's Loyalist corps at Kings Mountain in October 1780.[21] Soured by the poor performance of the Loyalist militia at Kings Mountain, Cornwallis halted organizing the militia until the loss of the light infantry at the Cowpens forced the resumption of recruiting efforts. By that

---

[18] Clay, "Staff Ride Guide to the Saratoga Campaign," 22.
[19] "An Introduction to North Carolina Loyalist Units," On-Line Institute for Advanced Loyalist Studies (website), accessed 7 November 2018.
[20] Babits and Howard, *Long, Obstinate, and Bloody*, 80.
[21] Wilma Dykeman, *With Fire and Sword: The Battle of Kings Mountain 1780* (Washington, DC: National Park Service, Department of the Interior, 1991), 12–14. The *overmountain* region was the area to the west of the Appalachian Mountains settled principally by Scots-Irish settlers after the French and Indian War, in defiance of English law. Although conscious of old abuses by the crown, and resentful of the high-handed behavior of the royal governor before the American Revolution, these western settlers did not become involved in the war until they were threatened by Loyalist incursions. Those Patriots who crossed back to the east over the mountain to fight against Loyalist forces were referred to as *Overmountain men*.

stage in the campaign, few Loyalists were willing to volunteer, partly out of fear of Patriot retribution but also due to frequent abuse by the British regulars, as Commissary Charles Stedman ruefully related: "[the British] could not have proceeded but for the personal exertions of the militia, who, with a zeal that did them infinite honour, rendered the most important services . . . in return for these exertions, the militia were maltreated by abusive language and even beaten."[22] Later in North Carolina, Cornwallis "erected the king's standard, and invited by proclamation all loyal subjects to repair to it, and take an active part in assisting him to restore order and constitutional government."[23] Ruin soon fell on the British efforts when a newly recruited Loyalist militia battalion was destroyed in a one-sided battle, Pyle's hacking match, in February 1781. The destruction of Colonel John Pyle's Loyalist unit made obvious the inability of the British to protect their own supporters, as noted by Patriot militia commander Colonel Andrew Pickens: "It has knocked up Toryism altogether in this part."[24]

The German Units

Besides recruiting from loyal Americans, Britain contracted for combat units from several German states, finding the payment easier than the effort to raise and equip additional regiments. The Landgraviate of Hesse-Cassel was the largest contributor to the British effort, and so Americans generically called all German units "Hessians" regardless of origin. The Germans were not truly mercenaries, as the regiments were organized for conventional European wars, and staffed with long-service professional soldiers who did not individually volunteer for overseas duty. The troops received no additional pay or incentives for being in America; instead, the financial benefits accrued to the German princes who contracted out their regiments. German regiments were organized differently than British or American units, with variations between different German states. Each German regiment was commanded by a colonel, seconded by a lieutenant colonel, major, and a staff of 18 other officers and enlisted. A typical German regiment was composed of five or six line infantry companies and one grenadier company. Each company was led by a captain, up to 3 lieutenants, and 10 NCOs and comprised between 114 and 165 enlisted. A German regiment could field between 525 and 690 muskets depending on how it was organized—surpassing the firepower of a British regiment, and on par with, if not surpassing, that of the Continental regiments.[25]

Two German units were among the reinforcements sent to Cornwallis's army in early 1781. The first was a single *jäger* (hunter) rifle company detached from its parent regiment and commanded by Captain Friedrich W. von Roeder. The company was authorized two officers, one NCO, and 78 riflemen. Jägers were fighters recruited from hunters and gamekeepers and trained and equipped to perform light infantry and sharpshooting missions. The second German unit was the *Von Bose Regiment* of five companies, which deployed to America in 1776. The *Von Bose* saw minor service in New York and New Jersey before taking part in the Savannah campaign in 1778. The *Von Bose* wore blue uniforms with white facings, similar enough to Continental uniforms to cause confusion on the battlefield. For the campaign, platoons of *Von Bose* infantry were task organized with platoons of the Ansbach jägers.[26]

WEAPONS
Muskets
The primary infantry weapon used by the British Army was the Short Land Pattern Brown Bess

---

[22] As quoted in John Buchanan, *The Road to Guilford Courthouse: The American Revolution in the Carolinas* (New York: Wiley, 1997), 244.
[23] Charles Stedman, *The History of the Origin, Progress, and Termination of the American War*, vol. 2 (London: J. Murray, 1794), 332.
[24] Buchanan, *The Road to Guilford Courthouse*, 39.
[25] Clay, "Staff Ride Guide to the Saratoga Campaign," 20–22.
[26] Babits and Howard, *Long, Obstinate, and Bloody*, 80–81, 88–89. *Ansbach jägers* were so called after the city of their origin: Ansbach, Germany.

.75-caliber smoothbore flintlock musket of varying lengths, which weighed around 10 pounds, fired a .69-caliber one-and-a-half-ounce lead ball, and mounted a deadly 16-inch socket bayonet. Tactics of the time emphasized the shock value of massed volley fire, so the weapons were not equipped with sights and soldiers seldom received marksmanship training. The British Army fielded the first Short Land Pattern muskets in the 1720s, and many were furnished to the colonial militia armories; consequently, many Brown Bess muskets saw service in the Patriot militia ranks. British fusilier regiments, such as the *23d Foot (Welch Fusiliers)*, were issued a fusil, a slightly lighter and shorter version of the Short Land Pattern musket.

At least 48,000 of the .69-caliber French Charleville Model 1763 and 1766 muskets were smuggled into the colonies from France beginning in 1777 and were adopted as the primary musket issued to Continental regiments. The M1766 musket was 57 inches long, weighed about 10 pounds, and fired a 1-ounce .65-caliber ball, which was often supplemented with the addition of lead buckshot in each paper cartridge (buck and ball) that would theoretically create a shotgun-like pattern of projectiles with each volley. Some American militia units carried the Charleville, but more carried Brown Bess muskets or a mix of personal rifles and shotguns.

The *Von Bose Regiment* was armed with .72-caliber Potzdam Model 1720 muskets, which had an overall length of 51 inches and a barrel length of 34 inches and weighed about 9 pounds. The .72-caliber bore meant the German troops could use the same .69-caliber cartridges used by the British, thus simplifying logistics. By 1780, the weapons fielded to the *Von Bose Regiment* were more than 50 years old and in poor repair, leading Major Johann Christian du Buy to unsuccessfully lobby for replacement with the Brown Bess muskets. Apparently, du Buy's request was ignored until after the Guilford Courthouse battle, when the regiment belatedly received new Brown Bess muskets from stores at Wilmington, North Carolina.[27]

Muskets versus Rifles

In terms of strict accuracy, the smoothbore musket had an effective range one-third less than that of a rifle—less than 100 meters versus 300 meters.[28] Yet, despite the obvious disadvantages in range and accuracy, muskets offered more advantages in combat compared to rifles, which is why muskets were the principal infantry weapon of European armies. First, mass-produced muskets were robustly made, designed for hard usage and ease of maintenance in the field. Second, smoothbores were much less affected by black powder fouling due to the loose fit of the ball in the barrel. When fired at close range, the heavy lead musket ball was capable of causing carnage in the enemy ranks, even more so with buck and ball cartridges. Finally, a bayonet-tipped musket gave a musketeer a decided advantage over a rifleman in close-in combat.[29]

Finely crafted muzzle-loaded flintlock rifles were employed by both sides, but in secondary or specialist roles. A skilled shooter using their own Pennsylvania rifle could accurately hit a squirrel at 200 meters and an adult-size target out to 300 meters. In exchange for such accuracy and range, the effective rate of fire was about one round per minute, as the rifleman had to measure powder from a horn, nest the ball into a greased patch, and pound the whole into the rifling with a ramrod—a task that increased in difficulty once unburnt powder fouled the bore. Rifles were designed for hunting, not combat; they were not designed to mount bayonets and would break apart if used in hand-to-hand combat. Variations in each handcrafted rifle meant a skilled gunsmith had to repair the weapons,

[27] Bill Ahern and Robert Nittolo, *British Military Long Arms in Colonial America* (Pittsburgh, PA: Dorrance, 2018), 409.

[28] Lawrence E. Babits, *A Devil of a Whipping: The Battle of Cowpens* (Chapel Hill: University of North Carolina Press, 1998), 13.

[29] Babits, *A Devil of a Whipping*, 13.

while the shooter had to cast their own bullets.[30] Pennsylvania or Deckard rifled flintlock muskets ranged in caliber from .36 to .48. The Ansbach jägers used a similar .67-caliber hunting rifle (*büchse*); thus, German riflemen were known as jäger, which means *hunter* in German. Smaller and shorter than a Pennsylvania rifle, the büchse has a correspondingly shorter effective range of 175 yards.[31]

Secondary Weapons

In terms of the tactics of the time, the bayonet was the mission-essential secondary weapon for regular infantry soldiers. After reducing the strength of an enemy's formation through firepower, an infantry regiment would march forward and break a shaken enemy line with the press of the bayonet. Bayonets for each army were all similar in basic design: a long spike blade varying between 14 and 18 inches in length and mounted to the muzzle of the weapon by a socket and stud.

Officers on both sides carried swords, both as a badge of rank and for close-range combat, and some specialist troops, such as the soldiers of the Ansbach jäger company, carried a short sword (*Hirschfänger* or deer catcher) when a bayonet was not practical. Continental officers were ordered by General Washington to carry a spear-like spontoon (a.k.a. half-pike) as a visible mark of authority on the battlefield. British officers had long rejected the use of spontoons on the battlefield, and some went further by exchanging their swords in favor of a privately procured fusil (light flintlock musket) and bayonet, thus reducing their visibility to a sharp-eyed Patriot rifleman. The practice was condemned by Lieutenant General Clinton, as he believed a fusil-armed officer was too easily distracted from their command duties:

> General Burgoyne and I have often represented the absurdity of officers' being armed with fusils, and the still greater impropriety . . . by which they neglected the opportunity of employing their divisions to advantage. . . . an inconvenience which I had long apprehended might result from officers' carrying fuzees, which was then and had been the general practice on the American service.[32]

One common secondary weapon carried by American soldiers was the tomahawk or hatchet that made for a lethal close-range weapon capable of killing or immobilizing an opponent with a heavy blow.

Dragoons

The cavalry organizations on both sides in the Southern Campaign were routinely task organized with light infantry to form robust mobile units suitable for screening, pursuit, and delaying actions in support of the main infantry force. Unlike European cavalry, who were only trained to fight while mounted, light dragoons were trained to fight both on horseback and as dismounted skirmishers with the light infantry. For dismounted work, dragoons carried a smoothbore flintlock carbine, while the heavy-bladed saber was the weapon of choice when mounted, as mounted shock action with the saber was preferred by commanders on both sides.[33]

Artillery

Nomenclature for artillery guns was based on the weight of the solid shot; a 3-pounder gun fired

[30] Babits, *A Devil of a Whipping*, 14–15.
[31] Babits and Howard, *Long, Obstinate, and Bloody*, 81.
[32] "British Officer Infantry Weaponry, 1768–1786," His Majesty's 62d Regiment of Foot (website), accessed 10 January 2018.
[33] Babits, *A Devil of a Whipping*, 20.

a solid shot weighing 3 pounds, and so on. Guns were made of durable bronze; 6-pounder guns mounted on a wood and iron two-wheeled carriage weighed 900 pounds and required two horses to move. Three-pounders mounted on a wheeled carriage weighed about 500 pounds, light enough for movement by a single horse or the gun crew during a battle. Three- and 6-pounders were considered field guns due to their relatively light weight and mobility. The standard crew for a 6-pounder field gun was 15 soldiers led by a commissioned officer and seconded by a sergeant and a corporal; the 3-pounder crew was a minimum of one officer, two gunners, and eight fighters. Each gun crew had one or more specialist gunners who calculated distance and elevation and rammed, aimed, and sponged the cannon, while the bombardier handled the vent and loading of the correct ammunition in the breech. The remaining crew consisted of matrosses and artillery privates, the soldiers who positioned the gun and passed ammunition to the bombardier.[34]

Cannons fired four major types of ammunition: shot, grape, canister, and shell. Solid cast iron round shot was used primarily against massed infantry and cavalry targets but was also used for battering fortifications and engaging in counterbattery fires. Maximum ranges for a 6-pounder gun firing solid shot was around 1,000 meters; the 3-pounder's range was about 800 meters. Grapeshot was a medium-range antipersonnel round consisting of a cluster of golf ball-size metal balls loaded in a wood and canvas container that disintegrated during firing to release a cluster of projectiles toward the target. Canister, or case shot, consisted of musket balls packed in a tin container that shattered on discharge to release a shotgun-like fan of bullets against enemy formations at ranges of less than 400 meters. Shells were hollow iron spheres filled with explosives, primarily fired at high angles from howitzers and mortars to explode on or within enemy fortifications and installations.[35]

TACTICS
Infantry Tactics

By 1780, conventional linear tactics built around the flintlock musket and bayonet had been in use for more than 100 years. For combat, infantry were deployed in linear formation, usually in two to three ranks to maximize the effect of en masse or volley fire to the front. Infantry regiments formed the standard infantry tactical unit, known as heavy infantry or the line infantry. The primary tactical goal of every army was the synchronized employment of all arms, infantry, dragoons, and artillery to break the enemy line of battle. Once the enemy line broke, a bayonet charge and pursuit by dragoons was employed to seal the victory for the still-intact infantry line.

The standard sequence of events for battle opened with an approach march in column formation by the attacking army to reach a suitable battle position. Open ground, with a natural obstacle such as a river or swamp to protect one or both flanks was considered ideal for an army assuming a defensive role. When possible, the attacking army would deploy from its marching columns into battle line out of range of the enemy's cannon systems. Once deployed, the attacking force would advance to within 100 yards of the enemy line, the effective engagement range of the musket. During this approach march, skirmishers, sharpshooters, and artillery on both sides would engage to attrite and demoralize the enemy force.

Once within a suitable killing distance, infantry regiments would begin engaging with controlled volley fire directed by the officers, which was intended to shock and break cohesion of the enemy unit. A well-trained unit could load and fire its weapons about three times per minute while under fire;

---

[34] Janice E. McKenney, *The Organizational History of the Field Artillery, 1775–2003*, Army Lineage Series (Washington, DC: U.S. Army Center of Military History, 2007), 9–10.
[35] "Artillery," AmericanRevolution.org, accessed 26 June 2019.

in practice, volley fire by company or division was done to avoid having the entire regiment without loaded muskets. Speed was stressed over accuracy, as the shock of repeated volleys was intended to stagger and disorder the enemy line. Under ideal combat conditions, about 20 percent of the rounds from a volley were expected to hit an enemy infantry line at 50 yards. Casualties were naturally lower at greater engagement ranges or when the natural cover or fortifications were involved in the fighting. The cumulative effect of sustained volley fire was intended to produce casualties and a loss of cohesion that would soften up the enemy line sufficiently for a bayonet attack. Fierce hand-to-hand fighting could take place if the troops in the weaker line stood and fought; more likely, the disorganized side would break and retreat. Few infantry regiments, even well-trained ones, possessed the discipline to actually stand and receive a bayonet charge, and for the poorly equipped Patriot militia and state units, retreat was the usual outcome when facing British regulars.

In addition to the line (heavy) infantry, there were a number of specialist infantry units. Light infantry companies were specially trained to advance in an open skirmish line, performing screening and reconnaissance duties well ahead of the parent regiment. Skirmishers operated in fireteams of two to four fighters, with one or more soldiers engaging the enemy with aimed fire while the others reloaded, all well-dispersed to present a reduced target to enemy skirmishers. During the Southern Campaign, the Americans generally used their light infantry in their designated role, while their British counterparts often consolidated the light units into a provisional line infantry battalion. British and Hessian grenadiers, for which there was no American equivalent, were generally consolidated into provisional battalions for concentrated employment as specialist assault troops. Rifle-armed troops on both sides were similarly employed as light infantry, with tactics modified to account for their slower rates of fire and vulnerability to line infantry. Besides performing skirmishing tasks, riflemen were commonly used to engage enemy commanders and weapon crews, usually from the flanks or an overwatch position where reach and accuracy of the rifles could be used to maximum effect. Shortly before the Guilford Courthouse battle, Major General Greene task organized his rifle battalions, light infantry companies, and dragoons into what he termed a *corps of observation*—a combined arms grouping that maximized the effectiveness of each system while minimizing individual vulnerabilities. In particular, the riflemen needed the close presence of bayonet-armed light infantry and friendly dragoons for protection against hostile infantry and dragoons. With the notable exception of the Kings Mountain battle, every major battle in the South was fought by musket-armed Continental and British regulars using modified linear tactics, while dragoons, riflemen, and artillerymen performed important but auxiliary roles.

Dragoon Tactics

In theory, dragoons were organized and equipped to ride to their place of battle, dismount at a distance from the objective, and maneuver into combat fighting with muskets or rifles. These tactics required the dragoons to leave the horses under the care of every fourth person in the unit. As this practice would considerably reduce the firepower of the squadron, the tactic was seldom used in combat. Instead, dragoons fought mounted, attacking the flanks or rear of vulnerable infantry units with slashing broadsword attacks, while relying on attached infantry to provide a base of fire and protection against a superior force. Other battlefield tasks for the dragoons were the engagement of hostile mounted troops, reconnaissance, and pursuit of a defeated enemy. At the Battle of Guilford Courthouse campaign, Lieutenant Colonel Tarleton's *British Legion* consisted of 272 mounted troops, while Greene's army fielded 244 dragoons divided between Lieutenant Colonel Henry "Light-Horse Harry" Lee and Washington's Corps of Observation.

Artillery Tactics

In terms of battlefield employment, infantry and dragoon tactics were relatively simple and straight-forward. By contrast, the artillery arm was a specialized, technically oriented branch requiring skilled officers and soldiers to function properly on the battlefield. Each type of cannon (gun) had peculiarities and variables impacting the weapon's effectiveness, factors compounded by the effects of wind, temperature, and humidity. Consequently, artillery specialists required mathematical skills and a great deal of training before gaining proficiency as gunners. As casualties were inevitable in combat, cross-training each fighter to learn the tasks of the others was necessary. Once crews gained proficiency, the company commander would train the gun sections to function together. Tactically, sections of two to three guns were commonly assigned in a direct support role to an infantry brigade, although the army commander might elect to keep them in general support, ready for counterbattery, harassment, or reinforcement roles. Gunners preferred to emplace their weapons well out of enemy rifle range, positioned on the friendly flank so that the guns could enfilade (fire down the long axis) of the enemy infantry as it closed within range. For long-range engagements, gunners calculated the fall of the shot to create a bounce or ricochet effect through the enemy line, thus maximizing casualties. At engagement ranges less than 400 meters, gunners would switch from solid shot to grape or canister and would continue the engagement until the enemy line was within 100 meters, too close to safely fire. The guns could then revert to a flank protection role, although they were often withdrawn into reserve to avoid casualties from enemy rifle fire. On the attack, guns were unlimbered (detached or unpacked) from their vulnerable draft animals, and the crews pushed and dragged the guns forward, ready to lend immediate fire support at the quick halt. In improved defensive positions, guns were emplaced on the flank of the principal infantry line, positioned behind a redoubt of earth in such a way to enfilade the enemy infantry with grape or canister rounds before reaching the friendly works —without being engaged in return by enemy riflemen or guns. The longer-range 6-pounders were emplaced to perform both antipersonnel and counterbattery tasks against hostile artillery.[36]

Fortification and Defensive Tactics

The use of deliberate fortifications during the American Revolution was fairly common in the Northern theater but seldom used in the Southern Campaign. This discussion will focus principally on temporary field works, generally a trench line with a sloped earth curtain, redans, redoubts and lunettes (see selected glossary). When time permitted, more elaborate features like abatis (obstacle of felled trees with sharpened branches facing the enemy) and chevaux-de-frise (a defense of timber or iron barrel with spikes and often strung with barbed wire) were added to slow and canalize enemy assault columns. When properly integrated together, field works were intended to protect friendly troops and artillery from enemy fire, while also hindering the enemy's ability to penetrate into the depth of the defenses. When possible, field works incorporated favorable terrain and were laid out for mutual support with interlocking fields of fire. Especially when reinforced with fascines (rough bundle used to strengthen a structure) and gabions (similar to modern Hesco barriers), earthen fortifications could absorb considerable bombardment before collapse, allowing the defensive force to inflict disproportionally heavy casualties on an attacking infantry force.

---

[36] McKenney, *The Organizational History of the Field Artillery, 1775–2003*, 11–12.

Tactics in the Southern Campaign

The tactical offensive was the preferred form of operations during the American Revolution, and commanders would generally assume a defensive posture for economy of force reasons, either an area defense to deny the enemy access to terrain, or in a point defense role to protect an installation or town. Tactics were often unconventional, dictated by the types of units and equipment and logistics support available, and further influenced by physical and human terrain in the Carolinas:

1. British units abandoned the conventional three-line deployment in favor of a two-line combat deployment used by the Continental Army, which allowed the regiment to cover a greater front.
2. Regiments were deployed in looser formations, with more space between individual soldiers, in adaptation to the denser terrain of the South.
3. Due to their chronic shortage of heavy infantry, British light infantry and grenadier were often employed as conventional infantry.
4. Field guns were not often used in the field; when available, they were kept in general support under the direct command of the senior commander on the field.
5. Riflemen were used to attrite enemy commanders and crew-served weapon systems.
6. Riflemen, dragoons, and infantry were routinely task organized for mutual support.
7. Out of necessity, the American army had to use militia and state troops in an offensive role.
8. Greene maintained a strategic defensive posture but assumed a tactical offensive posture through local attacks to keep the British off-balance. Cornwallis followed the "best defense is a good offense" approach and would aggressively attack any Continental unit within striking distance of his regulars.

LOGISTICS

Each side operated with a significant set of logistics constraints that shaped the course of the campaign. At the strategic level, Great Britain had a well-organized system to move supplies and ordnance stores from England to ports in South Carolina. At the operational and tactical level, frictions and inefficiencies (in the form of poor roads and insurgent attacks) hindered the British distribution of supplies and reinforcements from ports to forward operating bases (FOBs) like Ninety Six and Camden. Frustrated with his lumbering supply trains during his pursuit of Greene, Cornwallis burned all nonessential wagons and supplies and attempted to live off the land—a decision that led his troops and horses to the brink of starvation in the weeks before the Battle of Guilford Courthouse. Afterward, Cornwallis bitterly realized his disregard for sound logistics arrangements had forced his command beyond its logistical culmination point. As a consequence, he had to cede the initiative to Greene after Guilford Courthouse and withdraw his command to refit—leaving Greene's army effectively in control of all but the coastal ports of the Carolinas. By contrast, Greene's close attention to logistics matters, both friendly and enemy, ultimately paved the way to a decisive strategic victory in the Southern Campaign. Thus, a good understanding of the logistical systems and methods used by both sides is vital to understanding why the Southern Campaign unfolded as it did.

Strategic Logistics

At the beginning of the war, the Americans were at a severe disadvantage to Great Britain, with Congress and the state governments having to cobble together a national logistics system from scratch. The Second Continental Congress had the notional responsibility for arming, equipping, and supplying Continental forces in the field. In reality, Congress had little financial or political power to dictate logistics priorities to the states, so most logistics support for American units came from the state governments. Congress took the first step to reform the Continental Army logistics system by creating a commissary general, quartermaster general, and commissary of artillery. Primarily concerned with strategic logistics, this embryonic Continental staff also provided assistance to their counterparts in the geographical departments and field armies. In terms of a strategic industrial base in America, most goods of military value were produced by local, private entrepreneurs. Any controls or priorities placed on these businesses were imposed by state governments, which were also principally responsible for raising and equipping their own militia organizations and Continental Army regiments recruited from within the state. Many of the various types of supply —food, for example—had to be sought out and purchased directly from the producer by a purchasing agent. Both the states and the Continental departmental quartermasters employed purchasing agents to acquire goods and materials to support their respective units, creating competition for scarce goods that inflated prices. Speculation and fraud ran rampant, and much time and energy was spent by government officials attempting to curb these problems. Adding to these headaches was the fact that Continental paper money, not backed by a system of taxation or currency reserves, was rapidly devalued, which made the challenge of acquiring supplies even more problematic. By 1781, American quartermasters lacked money and specie, and so had to impress provisions and livestock from private owners, in exchange for a receipt for later payment—but only if the owner could prove their loyalty. Effective in the short term to alleviate supply shortages, such strong-arm tactics carried the risk of damaging civilian support for the Continental war effort.

A particularly acute problem for the rebellious colonies was obtaining weapons and ordnance stores, as no large-scale manufacturing capability existed in America. Working through a network of secret agents, Congress gained covert support from the French and Spanish governments, who saw the opportunity to hurt the English by extending support to the rebels. The trickle of smuggled aid turned into large shipments of war materiel after the American victory at Saratoga in 1777, which convinced France to declare war on England. By the end of the war, more than 100,000 muskets and bayonets, more than 200 cannon, and many tons of ordnance stores had been delivered to the Americans from French and Spanish arsenals—all purchased on credit extended by the French monarchy.[37]

Compared to that of the new American nation, the British strategic logistical system was well established, roles and responsibilities were delineated, and the whole was backed by a robust financing system. That did not mean the system was efficient, but by almost any measure the British system was significantly better than the American system. On the plus side, Britain had more than a century of experience mounting expeditionary operations and had a well-developed network of arsenals, factories, and depots capable of producing large quantities of supplies and equipment to sustain an expeditionary army in North America. Furthermore, the Royal Navy was powerful enough to ensure the movement of supplies to the British Army's overseas field forces would at least reach the North American seaports without serious interference by enemy vessels.

To oversee provisioning in North America, the Treasury Board commissioned a civilian commissary general of provisions to work for each commander in chief. The commissary general was seconded by a military deputy and supported by a small staff of civilian employees. The commissary general department had the responsibility to inspect and verify delivery of provisions for troops and

<hr>

[37] Larrie D. Ferreiro, *Brothers at Arms: American Independence and the Men of France and Spain Who Saved It* (New York: Knopf, 2016), 335.

livestock. Duties ranged from overseeing contractors, inspecting and verifying delivery of goods, and settling accounts with vendors. Although subject to directives from the Treasury Board, the commissary general for America received their day-to-day orders from the commander in chief, Lieutenant General Clinton.

The logistics process began in the British Isles, where provisions and military stores ordered by the commissary general were collected from factories, depots, and suppliers and loaded on chartered merchant ships at the Irish port of Cork. After a convoy across the Atlantic Ocean, the transports unloaded at regional depots, principally Charleston, South Carolina, and Savannah, Georgia. Every year during the war, some 400 ships were required to move the necessary supplies and reinforcements from Great Britain. Assuming the supplies shipped from England were not spoiled or destroyed in transit, port quartermasters had to receive, store, and distribute the supplies across a considerable distance on unimproved roads, all while under constant threat of rebel attack. As the British did not have a dedicated quartermaster corps to manage the entire supply chain, civilian agents coordinated the purchase of transportation, storage, and distribution facilities, while accounting and safeguarding thousands of pounds worth of property and supplies.

Ordnance stores, quartermaster supplies, and replacement equipment and uniforms only formed a small portion of the cargo shipped from Great Britain. Instead, the largest portion of the cargo consisted of consumable items for the soldiers: principally pork and beef, wheat flour and hard bread, and butter and salt, but also including other foodstuffs like oatmeal, peas, cheese, bacon, fish, raisins, and molasses. Scurvy, brought on by a lack of vitamin C, was a perennial problem, especially during the winter months, so soldiers were given large quantities of spruce beer, vinegar, and sauerkraut. When in season, fresh vegetables were shipped as well, but their issuance was intended only for hospital patients, as healthy soldiers were expected to obtain or raise their own produce.

British Operational and Tactical Logistics

Following the successful capture of Charleston in May 1780, Clinton designated Commissary General Nisbet Balfour as the military commandant for the Charleston port and garrison. Besides directing the defenses of the port, Balfour was responsible for safely distributing supplies and reinforcements to field magazines at Camden and Ninety Six in South Carolina and the base at Augusta, Georgia. Balfour was assisted by a commissary of captures, who safeguarded Loyalist property in addition to collecting and repurposing provisions, cattle, and property seized from the king's enemies. Repeated insurgent attacks on the British lines of communication across South Carolina required the diversion of many regular troops for security tasks. Finally, to ensure Cornwallis's army had sufficient supplies to support the next phase of the offensive, British quartermasters had to carefully synchronize the stockpiling of supplies to coincide with local harvests.

> It was about this time [June 1780] that Cornwallis changed the instructions. . . . He now considered it ill advised to march his army through North Carolina before the harvest, and took strong measures to induce impatient [Loyalist] partisans not to rise until after the crops had been gathered. . . . the work of supplying the base at Camden with salt, rum, regimental stores, arms, and ammunition was under way, so that a further advance of the army beyond that point would be safeguarded. Due to the distance of transportation and the excessive heat of the season, the work was one of infinite labor, requiring considerable time.[38]

---

[38] *Historical Statements Concerning the Battle of Kings Mountain and the Battle of the Cowpens* (Washington, DC: Government Printing Office, 1928), 9–11.

Cornwallis had a small staff to help in planning and coordinating logistics support. Major Richard England of the *47th (Lancashire) Regiment of Foot* served as the quartermaster general, while Colonel Charles Stedman, a Philadelphia Loyalist officer with prior field service, was appointed as commissary general. When putting together a logistics plan, the British staff officers worked with Commissary General Balfour to forecast, order, and coordinate the delivery of bread, vegetables, and meat from the forward British depots to the units in the field by contracted wagons. Replenishment of supplies at Camden was fairly straightforward, as quartermasters could use boats to ship supplies up the Santee River. However, Camden lay on the edge of the fall line, the point where the Santee was no longer easily navigable; furthermore, streams in the up-country generally ran perpendicular to Cornwallis's planned line of march. As a consequence, Cornwallis could not use cheap and reliable water transport to sustain his movement farther into the interior of the Carolinas. Instead, the farther his army marched from Camden, the longer his line of communications grew, so the British units would grow increasingly dependent on provisions drawn from the countryside.

Besides consumables for the soldiers, procurement of forage for the draft animals and cavalry horses was vital, as each animal needed about 50 pounds of green forage (or 25 pounds dry) per day. Forage was too bulky to haul from Charleston, so Cornwallis's logisticians had carefully timed the availability of fresh forage to support any major movements. For example, Cornwallis's delay in marching into North Carolina was designed as much to allow his soldiers and friendly Loyalists to gather in the forage crops as for the wheat and corn harvest. During the winter months, the army would have to periodically move from district to district once easily available forage in a particular area was exhausted.

American Operational and Tactical Logistics

Prior to assuming command of the Southern Department, Greene had learned firsthand the value of denying supplies to the enemy during his stint as the commissary general. Washington ordered Greene to take his troops and sweep the countryside clear of livestock and forage, a task at which Greene excelled. "You must forage the country naked," Greene told one commander, and to Washington he reported, "[The] inhabitants cry out and beset me from all quarters, but like Pharoh I harden my heart."[39] Greene's good performance in supplying the army during the Valley Forge winter led to his appointment as the quartermaster general of the Continental Army. Although Greene detested the role, he performed well and made significant improvements in transportation and distribution procedures; so when he assumed command of the Southern Department, he took swift measures to improve the logistics posture of the command. First, Greene appointed Lieutenant Colonel Edward Carrington as quartermaster general and Lieutenant Colonel William Richardson Davie as commissary general, with each given the authority to do the job properly. Second, Greene dispatched officers throughout the department to map out fords and bridges and assess the suitability of using waterways to efficiently move supplies and troops. The results of the survey convinced Greene to order the construction of a fleet of 100 *bateaux*, flat-bottomed boats suitable for ferrying supplies and troops across rivers. To better support the Continental Army in the field, Greene established a factory or logistics depot at Salisbury, North Carolina, to facilitate receipt and distribution of supplies, plus repair and manufacture of needed items. Before Greene decided to move his army away from Charlotte, North Carolina, in the fall of 1780, on account of the poor supply situation, Colonel Thaddeus Kosciuszko was dispatched to survey the Pee Dee River valley, paying particular attention to "the quantity of

---

[39] Terry Golway, *Washington's General: Nathanael Greene and the Triumph of the American Revolution* (New York: Henry Holt, 2006), 162.

Produce, number of Mills, and the water transportation that may be had up and down the River."[40] Greene's decision to split his army in late 1780 was driven largely by logistics considerations, presaging Napoleon's dictum that an army should march divided and fight united.[41]

After carefully studying the logistics situation in the Carolinas, Greene realized Cornwallis's army was uniquely vulnerable to interdiction of its local sources of provisions and fodder. Thus, Greene focused his efforts on turning the Carolinas into a literal food desert to deprive Cornwallis's army of necessary supplies. Greene repeatedly prodded the local Patriot guerrilla commanders, principally Thomas Sumter and Francis Marion, to attack vulnerable enemy supply convoys and magazines. When Brigadier General Morgan marched his troops west of the Catawba River, Greene's instructions on denying provisions to Cornwallis were explicit: "The object of this detachment is to give protection to that part of the country and spirit up the people—to annoy the enemy in that quarter—*collect the provisions and forage out of the way of the enemy*, which you will have formed into a number of small magazines, in or near the position you may think proper to take."[42] Greene's troops operating near Cheraw Hill followed the same instructions, and in short order British foraging expeditions had an increasingly difficult time finding provisions. When Cornwallis's army withdrew from Charlotte to Willingboro in late October 1780 after the Kings Mountain defeat, the two-week movement was plagued with shortages of meat and bread and the complete absence of rum. Moreover, Morgan's flying corps so thoroughly swept up supplies from the Catawba region that Tarleton's soldiers ran out of food the day before the Battle of Cowpens. The pinch of hunger certainly contributed to the collapse of the British infantry during the latter part of the fight.[43]

As Cornwallis's army marched into North Carolina in pursuit of Greene's withdrawing army, the lines of communication back to Camden were abandoned, and the British troops were completely dependent on drawing supplies from an already picked-over landscape. Shortages of wheat flour meant dried corn was the principal source of carbohydrates. A common practice by British foraging detachments was to encamp close to a mill, which made for convenient processing of any grain brought in by foraging detachments. The need to gather and process so much grain meant the British commanders had to devote an increasingly greater amount of time, energy, and personnel away from their primary mission and toward foraging, cutting wood for fires, grinding meal, and making bread—with the entire process repeated every third or fourth day. The whole foraging process retarded Cornwallis's movement so that his army could average no more than 10 miles (16 km) per day in their pursuit of Morgan. After camping at Ramsour's Mill to collect supplies, Cornwallis made a risky decision on 25 January 1781 to lighten his army by destroying all nonessential wagons and supplies: "I employed a halt of two days in collecting some flour, and in destroying superfluous baggage and all my waggons, except those loaded with hospital stores, salt, and ammunition, and four reserved empty in readiness for sick or wounded."[44] During the subsequent Race to the Dan, Cornwallis's lightened army moved much faster and several times came close to cutting off elements of Greene's army, but in the end, the risk Cornwallis assumed by burning his logistics trains was not enough to ensure a successful pursuit.[45] As Cornwallis's army remained in North Carolina trying to corner the Americans,

[40] Gen Nathanael Greene letter to Col Thaddeus Kosciuszko, 8 December 1780, Raab Collection, accessed 9 October 2019.

[41] Babits and Howard, *Long, Obstinate, and Bloody*, 9. The standard American ration mirrored that of the British Army: up to a pound of fresh bread and meat per day supplemented by smaller quantities of butter, cheese, oatmeal, rice, peas, and sauerkraut, along with condiments such as vinegar, sugar, and salt. Large quantities of spruce beer or porter was issued for antiscorbutic properties, while a daily gill of rum helped to take the edge off hard field service, especially in cold weather. In the field, the quartermaster substituted hard tack or biscuit if wheat flour and baking facilities were available. More commonly, cornmeal was substituted for wheat flour, eaten in the form of cornmeal mush or firecakes cooked over an open fire.

[42] Gen Greene to BGen Morgan, "The Detached Command Formed Which Won Cowpens, 16 December 1780," in Theodorus Bailey Myers, ed., *Cowpens Papers: Being Correspondence of General Morgan and the Prominent Actors* (Charleston, SC: News and Courier Book Press, 1881), 9, emphasis added.

[43] Babits, *A Devil of a Whipping*, 156–57.

[44] Earl Cornwallis, *An Answer to that Part of the Narrative of Lieutenant-General Sir Henry Clinton, K. B., Which Relates to the Conduct of Lieutenant-General Earl Cornwallis, during the Campaign in North-America, in the Year 1781* (London: J. Debrett, 1783), 25.

[45] The Dan River flows near the border between North Carolina and Virginia.

the lack of wagons to carry supplies and the lack of supplies in the region meant the British had to move more often and devote more time to gathering and processing provisions. By early March 1781, Cornwallis's troops were starving, living on little more than a single meal of four ounces of corn flour and four ounces of meat per day. The day of the Guilford Courthouse battle, the British troops arose before dawn and made an approach march of 15 miles (24 km) before fighting—all without an issuance of rations for the day.

By comparison, Greene's army had the benefit of local access to food and forage, as the American militia and light units were ruthlessly efficient at taking provisions and livestock from known or suspected Tory sympathizers. Furthermore, by operating in upper North Carolina, Greene's army benefitted from secure lines of communication to the Continental logistics depots. Once Greene's army reached sanctuary in Virginia, the American soldiers rested and replenished with "wholesome and abundant supplies of food in the rich and friendly county of Halifax."[46] In comparison to the wretched redcoats, Greene's troops were well fed before the battle, and those that survived the battle were rewarded with the issuance of two day's rations and a gill (quarter of a pint) of rum.

ENGINEER SUPPORT

In many respects, engineer operations during the Southern Campaign differed little in concept than modern mobility and countermobility processes. As in most other warfighting functions, the British Army had a distinct advantage: its engineer officers were formally trained. By contrast, the Americans were fully reliant on foreign-trained officers. Greene had Colonel Kosciuszko from Poland as his staff engineer, while Cornwallis had Lieutenant Henry Haldane of the Royal Engineers as his senior engineer officer. Both armies lacked specialist engineer or pioneer units, so necessary engineering tasks were done by ordinary soldiers under the direction of the staff engineer. At the Salisbury depot, Greene established what today would be termed a vertical engineering detachment, with carpenters capable of performing limited construction tasks. No record exists as to a similar organization in Cornwallis's field army.[47]

Mobility

Mobility operations are those engineer activities that mitigate the effects of natural and manmade obstacles to enable freedom of movement and maneuver for a combat force; namely, construction and improvement of roads and bridges and the reduction of enemy fortifications by sapping operations. The most notable example in the mobility domain was the Americans' bateaux train, which assured Greene's army of operational mobility during the Race to the Dan, regardless of the river levels. Construction of the boats, technically a quartermaster function, was accomplished by the troops of the artificer company under the supervision of Kosciuszko. Also included within the mobility domain was Greene's ordering of a detailed topographical survey of the roads, fords, and rivers in the region. Although he had never before visited the Carolinas, the resulting topographical information yielded for Greene an exceptionally good situational understanding of his battlespace. Greene's foresight and good Continental engineering staff resulted in a successful operational-level withdrawal across the Dan River without significant loss of personnel or materiel. Moreover, the topographical data allowed Greene to choose his battlefield at Guilford Courthouse due to its mix of natural obstacles and

---

[46] Henry Lee, *Memoirs of the War in the Southern Department of the United States* (Washington, DC: P. Force, 1827), 249.

[47] Gen Greene to Colonel Thaddeus Kosciuszko, 1 January 1781, in Dennis M. Conrad, et al., *The Papers of General Nathanael Greene*, vol. 7, *26 December 1780–29 March 1781* (Chapel Hill: University of North Carolina Press, 1994), 35.

good avenues of approach and withdrawal. By contrast, Cornwallis never seemed to have employed Lieutenant Haldane in a similar manner, even though Haldane's post-battle topographical map of the Guilford Courthouse battleground indicates he was a skilled engineer. During the campaign, Cornwallis made several bad tactical decisions on account of his lack of information about the road and water crossings, and the lack of suitable bridging material repeatedly delayed the British pursuit.

Countermobility

Countermobility operations include the use or improvement of natural and man-made obstacles to deny an adversary freedom of movement and maneuver. Due to the fluid nature of the Southern Campaign and lack of equipment and specialist engineer units, little work was done in the counter-mobility domain. The best example of countermobility operations was seen at the Catawba River crossings, where Morgan's soldiers piled logs and branches at the river fords to hinder the British infantry. A second, passive, countermobility measure was Greene ordering his militia to take up or destroy all boats within close proximity to the Yadkin and Dan Rivers, thus depriving the British the means to easily cross them. Finally, Greene chose the battleground at Guilford Courthouse because of the area's natural countermobility features, particularly the rugged belt of timber between the second and third lines, where today fallen trees provide good examples of natural countermobility obstacles. Yet, there are no indications from eyewitness accounts that Greene ever considered constructing obstacles to delay or canalize the enemy movements.

Survivability

Survivability operations are those military activities that alter the physical environment to provide or improve cover and concealment; generally speaking, this means the construction of fortifications to protect soldiers from direct fire (bullets and projectiles) and indirect fire (shells and shell fragments). As within the countermobility domain, little survivability engineering work was done during the campaign. When the Continental Army was preparing to cross the Dan River at Boyd's Ferry, Colonel Kosciuszko supervised a crew of militia and impressed slaves in erecting earthworks to guard the approaches. After the Battle of Guilford Courthouse, Greene ordered the digging of hasty fieldworks at Speedwell Furnace, North Carolina, to slow the expected pursuit by Cornwallis's army. In neither instance were the fieldworks tested in battle.

COMMUNICATIONS
Strategic Communications

Strategic communications between the Continental Congress, Continental commander in chief, and the American field armies tended to be much easier than it was for the British government. Continental armies usually operated on interior lines of communications and so face-to-face meetings and courier messaging tended to not only be much safer but also much faster. A courier message from Greene in the Southern Department could reach the governors of North Carolina or Virginia within a week, while a courier supported by relay of fresh horses could reach Philadelphia within two weeks, barring major delays due to weather or enemy action. Due to the long distances involved, Greene did not make personal visits to Congress or state governments after his trip south in the fall of 1780, instead relying on courier messages or liaison officers.

Strategic communications between British secretary of state for the colonies, Lord George Germain, Cornwallis, and Clinton could only travel as fast as the unpredictable Atlantic weather and currents allowed. Under the best of conditions, a journey from London to Charleston could take 60 to 90 days, although the trip was about two to three weeks shorter on the return voyage. Consequently, Germain had to cede effective strategic-level command and control to Clinton in favor of broad grand strategic guidance. Limited by the same time and distance factors, Clinton could only exercise a limited amount of operational control over Cornwallis, who gladly took advantage of the lack of supervision to craft his own campaign plans. As it was, the strategic communications between Germain and his commanders was not only slow but was largely ineffective due to poorly thought-out plans and directives.

Operational Communications

At the operational level, communications were passed primarily by courier. Given the fractious nature of the Southern Campaign, couriers needed an intimate knowledge of the region to avoid riding into hostile territory. To avoid notice, the couriers usually rode without escort and wore civilian attire—a risky venture, as they could be executed as a spy if caught. For exceptionally important messages, more than one courier was dispatched to better ensure that at least one of the letters (or verbal message) got through. Both sides also employed encoding systems to delay enemy exploitation of captured messages.

Tactical Communications

The courier system extended down to the tactical level as well, especially when elements were widely separated, such as Morgan's detached service from Greene prior to the Cowpens battle. When possible, commanders would give tactical orders verbally to subordinates while viewing the terrain in question. To help keep track of the orders, the adjutant would record the verbal orders in an order book and, time permitting, forward a copy to the commander in question. In Cornwallis's headquarters, Major John Despard of the *Volunteers of Ireland* performed duties as the adjutant general. Colonel Otho Holland Williams, in addition to his primary role as the commander of the Maryland Continental brigade, served as the adjutant general of the Southern Army.

During battle, flags, signal guns, bugles, fifes, whistles, and drums were all used in an attempt to transmit and receive timely information. Courtesy of Major General Friedrich von Steuben, Washington's inspector general, the Continental Army had a standardized system of signals based on drumbeats, flags, and music to coordinate tactical movement and maneuver on the battlefield. British commanders seemed to use whistles as a particularly effective means of signaling for simple movements. Combined operations with German and British units posed challenges due to possible language barriers. There is some evidence that French was used as a common third language for written communications between British and Hessian officers, although few other details exist as to exactly how orders between the two were processed.

INTELLIGENCE

Dragoon patrols were the principal means of gathering intelligence, supplemented by reports from scouts and spies and interrogation of prisoners and civilians. Besides serving as the eyes and ears of the commander, the dragoons also had an important screening mission, blocking hostile dragoons from gathering information on the main body of the army. During the period before the Guilford

Courthouse battle, Tarleton's dragoons were usually overmatched by the equally aggressive Continental dragoons, who were mounted on stronger and heavier Virginia-bred horses. Consequently, Cornwallis had little hard information about the location and composition of Greene's force in the days just prior to the Guilford Courthouse battle. When the British Army marched toward Greene's position on the morning of the battle, Cornwallis mistakenly thought Greene had doubled his actual strength of 4,400 troops. Moreover, the British tactical commanders lacked basic information about the composition and deployment of the Americans and the local terrain.

By comparison, Greene was a keen collector and user of intelligence and had taken great pains to learn all he could about his area of operations. During the Race to the Dan, Greene's intimate understanding of the battlespace gave the Americans a decided edge in outmaneuvering the enemy forces. Greene also chose his battle position at Guilford Courthouse only after a personal reconnaissance. Despite his longer presence in the Carolinas, Cornwallis seems never to have made a systematic effort to gain situational awareness and understanding of the area of operations. For example, Cornwallis had a faulty understanding of the road networks, hydrology, and topography of the Carolinas. During the pursuit of Morgan in January 1781, Cornwallis set his axis of advance toward supposedly better fording points on the upper Broad River, close to the Blue Ridge foothills. However, he failed to realize that the upper forks were more susceptible to wintertime floods, and the forward progress of the British force was greatly hindered. He noted that "the swelling of numberless creeks in our way, rendered all of our efforts fruitless."[48]

In the days before motorization and improved roads, soldiers had to pay close attention to the weather and assess its impact on military operations. Weather forecasting consisted largely of direct observation, looking for signs of changes in the wind direction and clouds. Precipitation impacted the movement of the armies during the run-up to the Guilford Courthouse battle, settling dust that would give away the movement of troops but in turn muddying fields and roads, which slowed the initial deployment of the British Army into battle lines. Similarly, the rain-swollen Richland Creek that ran across the front of the American third line presented a strip of slow-go terrain to hinder the British movement to contact. Lastly, high moisture was a major concern for soldiers carrying flintlock weapons, as keeping powder and cartridges dry was vital to reliable ignition in combat.

MEDICAL

In general, there were three kinds of hospitals employed by the Continental Army during the war: general, flying, and regimental hospitals. Each Continental department was authorized one or more general hospitals to provide long-term medical care and were normally located in permanent buildings far from the combat zone. The first American department hospital was established in 1776 at Charleston, South Carolina, where it remained until the end of the siege in May 1780. Afterward, a new ad hoc field general hospital was created at Charlotte, North Carolina, supported by two intermediate field hospitals at Hillsborough and Salisbury, North Carolina.[49]

When Greene took command, he was immediately dissatisfied with the poor administrative skills of Dr. James Brown, the chief surgeon of the department, and Greene found the general hospital "shocking to humanity."[50] Consequently, Greene ordered the general hospital moved to Salisbury in mid-December 1780 for closer supervision, after which Greene expressed his concerns to General

---

[48] Charles Ross, ed., *Correspondence of Charles, First Marquis Cornwallis*, vol. 1 (London: John Murray, 1859), 503.
[49] LtCol Louis C. Duncan, USA (Ret), *Medical Men in the American Revolution, 1775–1783* (Carlisle Barracks, PA: U.S. Army Medical Field Service School, 1931), 312.
[50] Duncan, *Medical Men in the American Revolution, 1775–1783*, 171.

Washington: "I will not pain your Excellency with further accounts of the sufferings of this army; But I am not without great apprehension of its entire dissolution unless the Commissary's and Quarter Master departments can be rendered more competent to the demands of the service. Nor is the clothing and hospital departments upon a better footing."[51]

When Greene's army withdrew from North Carolina in January 1781, the general hospital at Salisbury was packed up and evacuated into Virginia, which left the Americans dependent on flying and regimental hospitals. As the name implies, flying hospitals were mobile and would set up in any available civilian structure or in tents, if necessary. The facilities had only a few beds and an area set aside for surgery. Regimental hospitals, operated by regimental surgeons, were analogous to modern aid stations and were designed to provide immediate medical care in close proximity to the battlefield. Once the surgeon was finished with his treatment, the patient was transferred to a flying hospital for further treatment or to a general hospital for long-term care.

Infectious diseases were a particular concern of both armies, as large numbers of fatigued and malnourished soldiers living in constant close quarters were peculiarly vulnerable to epidemics. If officers and NCOs did not closely monitor camp discipline, particularly concerning proper disposal of body wastes in sinks (field latrines), outbreaks of diarrhea, dysentery, and typhoid fever were bound to occur. Smallpox was another concern, as one in three individuals contracted the disease and those who survived required an extensive convalescent period before returning to duty. The Continental Army began a program of smallpox inoculations early in the war, so by 1781 few instances were seen in the regular ranks. However, the disease was prevalent in the ranks of the American militia and British soldiers, and periodic outbreaks occurred in both armies during the winter of 1780–81. The greatest disease danger, especially for northern European soldiers with no natural immunity, came from mosquito-borne malaria and yellow fever. As Cornwallis described to Clinton, "This climate (except in Charleston) is so bad within one hundred miles [161 km] of the coast, from the end of June until the middle of October, that troops could not be stationed among them (sic) during the period, without a certainty of their being rendered useless for some time, for military service; if not entirely lost."[52] Even by late September 1780, the illness rates in the British ranks were so bad that Cornwallis ordered most of his infantry regiments north into the slightly cooler region of upper South Carolina. Cornwallis was ill during much of October, and so many of his troops were ill that he decided not to pursue the victorious Patriots after the Kings Mountain battle.[53]

Medical support for both armies after the Guilford Courthouse battle was generally poor, not through lack of effort but due to basic knowledge and training. Most American surgeons learned their trade by serving an apprenticeship with a surgeon, with a handful having received formal medical education at a university in Europe, like their British surgeon counterparts. Consequently, treatment of battlefield wounds was limited to bandaging wounds, probing for musket balls, and amputating mangled limbs. Surgeons were ignorant of the causes and prevention techniques for diseases, so surgical instruments, towels, and bedding were reused without cleaning. As a consequence, many soldiers who had survived wounding on the battlefield and the trauma of surgery were still more likely to die from a secondary infection than from the injury.

On the British side, a Dr. John M. Hayes was appointed physician general and inspector general of hospitals of the British Army in the Carolinas. Cornwallis apparently did not appoint a supervising surgeon for his field army, so the duties of setting up the field hospitals would have fallen on the regimental surgeons. During the campaign, the closest general hospital for the British Army would

[51] Duncan, *Medical Men in the American Revolution, 1775–1783*, 324.
[52] Duncan, *Medical Men in the American Revolution, 1775–1783*, 312.
[53] John R. McNeill, "How the Lowly Mosquito Helped America Win Independence," Smithsonian.com, 15 June 2016.

have been the one at Camden supervised by Dr. Hayes. While on campaign, the regimental surgeons, assisted by surgeon's mates, would have established field hospitals when the army camped for longer periods. During the early stages of the campaign, those soldiers too sick or injured to fight were evacuated by wagon train to a general hospital at Camden, then on to Charleston. A percentage of the wagons in Cornwallis's logistics train were designated for medical support, transporting the sick and wounded, and hospital stores. These medical vehicles were deemed important enough to escape destruction at Ramsour's Mill on 25 January 1781. But once Cornwallis cut his lines of communication back to Camden, the regimental surgeons had no choice but to carry the wounded and sick forward with the army. By the end of the Race to the Dan, the British surgeons had large numbers of sick troops in their wagons, with the closest general hospital and source of medical supplies at Camden 240 miles (386 km) to the south.

When the American army withdrew from the Guilford Courthouse battlefield on the afternoon of 15 March 1780, Greene ordered the dead and nonambulatory wounded to be left on the battlefield. The British regimental surgeons established field hospitals at the nearby Hoskins farmstead house and the New Garden Meeting House, and during the night and next day soldiers and local Quakers combed the battlefield, separating the dead for burial in nearby pits and removing the moveable wounded to hospitals. The British wounded were taken to New Garden Meeting House and the Hoskins house, while the American wounded were deposited at Guilford Courthouse. The local Quakers were generous enough to supply livestock, eggs, flour, and milk for nourishment, but the supply-starved British surgeons had few medicinal stores left for their own people, much less the enemy wounded. The day after the battle, Cornwallis wrote to Greene, requesting a cease-fire and the return of American surgeons to help with the large numbers of wounded. Cornwallis's proposal was closed with a plea for "a Supply of Necessaries & Provisions." Greene accepted the cease-fire and for two days, British and American surgeons jointly treated the wounded. When Cornwallis's army marched from New Garden Meeting House toward Cross Creek on 19 March 1781, it was followed by a convoy of 17 wagons filled with wounded and sick troops, with the most seriously wounded left in the care of the Tory-friendly Quakers.[54] In all, Cornwallis's army of 2,550 troops had suffered a 27-percent casualty rate at Guilford Courthouse: 96 dead, 413 wounded (more than 50 of them mortally), and 26 missing. In his official return of the battle, Greene reported a combat strength of 4,400 fighters, with 79 dead, 184 wounded, and 1,046 missing (mostly deserted militiamen), although modern scholarship pegs the casualties at 94 killed and 220 wounded.[55]

---

[54] Buchanan, *The Road to Guilford Courthouse*, 381–82.
[55] Babits and Howard, *Long, Obstinate, and Bloody*, 175–76.

PART II

To fully understand the strategic and operational dynamics leading to the Battle of Guilford Courthouse, a brief historical review of the American Revolution is in order. After the outbreak of the revolution in 1775, the British laid plans to reestablish order and loyalty in the rebellious colonies. Believing that the center of Patriot resistance originated in the New England area, Lord Frederick North (principal military advisor to King George III) devised a strategy of divide and conquer. First, isolate the Americans with a close naval blockade. Second, split the Northeast from the middle colonies by securing the Hudson River valley. Third, isolate the grain producing states of Pennsylvania and Maryland. Last, secure the Carolinas by seizing Charleston and Savannah, Georgia. The American commander in chief, General George Washington, adopted a strategy of attrition from sheer necessity, fighting only when necessary and making all efforts to preserve what he saw as the Americans' center of gravity—the regular or Continental Army. By doing so, Washington hoped not only to exhaust the British but to encourage intervention in the war by one of England's traditional enemies. Despite a promising start with several battlefield victories early in the war, British efforts to coordinate strategy and operational plans foundered due to bitter rivalries between senior leaders. One of the worst examples occurred in 1777, when General Sir William Howe focused his efforts to capture the strategically meaningless enemy capital at Philadelphia, instead of cooperating with General John Burgoyne's army marching south from Canada via the Hudson River valley. Howe's refusal to assist Burgoyne, despite expectations from London, left Burgoyne's army overextended and unsupported and ultimately vulnerable to encirclement by American forces. As a consequence, Burgoyne's force was cut off and forced to surrender at Saratoga, New York, in October 1777.[1]

THE BRITISH SOUTHERN CAMPAIGN

The American victory at Saratoga paved the way to open French involvement in the war, with a consequent loss of strategic freedom to maneuver. The American victory at Monmouth Court House, New Jersey, on 28 June 1778 showed conclusively that Britain was no longer capable of beating the resurgent Continental Army. Hoping to regain the initiative in North America, Lord North and Lieutenant General Sir Henry Clinton, the overall commander in chief of the British Army in America, turned their thoughts back to the South. South Carolina loomed large in their strategic calculus for both military and economic reasons. Economically, British control of Carolina rice and indigo crops would help to pay for war costs. Militarily, the Royal Navy could use the protected anchorage of Charleston Harbor to support their Caribbean operations, while the army could use South Carolina as a springboard for operations toward the Chesapeake Bay. Diplomatically, uncontested control of the Carolinas would strengthen the negotiating position of the crown for eventual peace talks. Tory expatriates in London promised a strong showing of Loyalist support at the reappearance of British authority. By all appearances, a campaign in the Carolinas appeared to be the type of low-risk, high-payoff operation the British sorely needed to regain the initiative in the war. Accordingly, Clinton and North settled on a new plan to secure the Carolinas. By leveraging support from the Royal Navy, Clinton could quickly project combat power from New York into the South before the Continentals could react. Once Savannah and Charleston were secured, British regulars would establish a strong outpost line along the Savannah River to block Continental incursions from Virginia. While the British mobile elements mopped up any insurgents in the area, trainers drawn from regular units

---

[1] David K. Wilson, *The Southern Strategy: Britain's Conquest of South Carolina and Georgia, 1775–1780* (Columbia: University of South Carolina Press, 2005), 59.

would work to organize and train an effective Loyalist militia. The key assumption underpinning the entire British campaign plan was the use of well-trained Loyalist militia to successfully pacify and secure South Carolina. Absent that, Lieutenant General Clinton simply did not have enough manpower in North America to secure both New York and South Carolina.[2]

Besides the rational strategic calculations, Clinton had a personal motivation in leading a Southern Campaign: revenge. Paired with naval commodore Sir Peter Parker, Clinton had been deeply embarrassed by his part in a joint army-navy operation that utterly failed to take Charleston in June–July 1776, an enterprise marred with lack of cooperation, misunderstanding, and timidity. The failure at Charleston came shortly after the American Declaration of Independence and served notice to all involved that the Americans planned to fight hard for their freedom. Consequently, American morale in the South soared, while British hopes of a quick end to the rebellion were dashed. Besides damaging his reputation within the British Army and royal establishment, Clinton developed a deep-seated mistrust of Royal Navy officers. In 1780, Clinton managed to put aside his misgivings and drew up an impressive campaign plan to capture Charleston by an indirect approach. Clinton made good use of Captain George Keith Elphinstone, a British naval officer assigned to command the naval elements supporting the army's sea landing and river crossings. Elphinstone ably served as a liaison officer between Clinton and Commodore Mariot Arbuthnot, so that the Charleston campaign would resemble a truly joint army-navy operation. Clinton's planning was done in secret, so the British were able to gain the strategic initiative from the Americans, who failed to recognize the shift in British strategy.[3]

The shaping phase of Clinton's campaign opened during the winter of 1778–79, when a British expedition from St. Augustine, Florida, easily captured the vital river port of Savannah and secured much of the Savannah River line in the winter of 1778–79. Shocked by the loss of Georgia, Congress dispatched Major General Benjamin Lincoln with a contingent of North and South Carolina Continentals to Charleston to organize a Southern Army with hastily mobilized militia units. Disaster soon befell the Americans when a combined Franco-American assault failed to retake Savannah in October 1779, thus leaving Charleston uncovered and vulnerable to an amphibious assault. Consequently, Clinton's assault force of approximately 8,000 infantry, plus some 1,500 dismounted dragooons, landed without incident to the east of Charleston Harbor on 11 February 1780. Learning from his previous failures, Clinton's plan—ably abetted by Captain Elphinstone—was carefully thought out and well-coordinated with Commodore Arbuthnot. Despite instructions from Washington to not carelessly risk the Continentals, General Lincoln allowed local political leaders to constrain his freedom of maneuver, and British troops easily cut off all land and river communications with the outside world. After a siege lasting several weeks, Lincoln surrendered the city and his army of 5,446 troops on 12 May 1780, the largest surrender of American arms during the war. The Continental soldiers were sent to prison ships, with the Continental officers and the Patriot militiamen paroled and released to their homes. Taking Charleston was the single greatest victory of the British for the entire war, and at that time the Carolinas seemed all but secured.[4]

CONSOLIDATION AND PACIFICATION

During the consolidation phase after the fall of Charleston, the *British Legion*, a mixed infantry-cavalry force commanded by Lieutenant Colonel Banastre Tarleton, overtook and wiped out a withdrawing Continental regiment at the Waxhaws along South Carolina's piedmont border with North Carolina.

---

[2] Robert Stansbury Lambert, *South Carolina Loyalists in the American Revolution* (Columbia: University of South Carolina Press, 1987), 79–81.
[3] Andrew D. Dauphinee, "Lord Charles Cornwallis and the Loyalists: A Study in British Pacification during the American Revolution, 1775–1781" (master's thesis, Temple University, 2011), 15–16.
[4] John W. Gordon, *South Carolina and the American Revolution: A Battlefield History* (Columbia: University of South Carolina Press, 2003), 84–85.

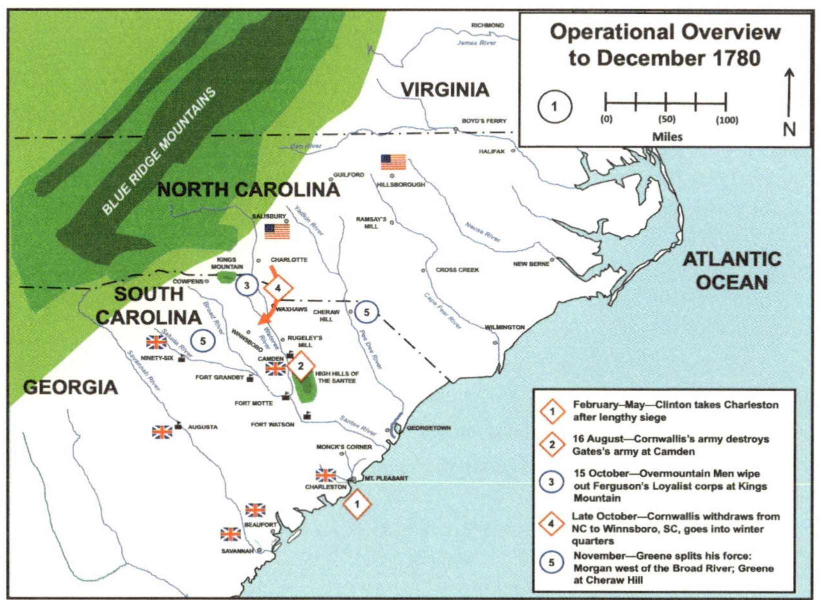

Map 1
Chuck Collins, Army University Press, and Harold Allen Skinner Jr.

Tactically, the Waxhaws capped off the seemingly triumphant British campaign in the South. When viewed within the context of the British pacification strategy, Tarleton's performance at the Waxhaws produced an information operations disaster. Whether true or not, accounts of brutal bayonetting of defenseless Continentals stiffened American resistance to British recruiting and pacification efforts. Lieutenant General Clinton added fuel to the fire by issuing a couple of poorly worded proclamations. The first offered a full pardon to ex-rebels in exchange for an oath of allegiance to the crown, lenient treatment resented by Loyalists, who expected harsh treatment for their traitorous neighbors. Clinton's second proclamation demanded oaths of loyalty to the king and mandated active service in the Loyalist militia from all men living within British-controlled territory. Clinton's proclamation handed the American cause another easy information operations victory, as many of those militia released after the capture of Charleston no longer felt legally bound by their paroles and returned to active resistance.[5]

In June 1780, Clinton and a sizable portion of the field army returned to New York to reinforce its garrison. Left behind was Lieutenant General Cornwallis and a contingent of 8,000 regulars and provincial troops to maintain base and line of communication security. Major Patrick Ferguson of the

---

⁵ Gordon, *South Carolina and the American Revolution*, 87–88.

71st Regiment of Foot (Highlanders) was appointed inspector of militia and tasked to organize and train a Tory militia regiment in each district to take over security duties at key forward bases like Ninety Six, South Carolina. Cornwallis originally expected Ferguson to have completed his organizational tasks in time for the spring campaign season, thus freeing Cornwallis to march into North Carolina and to repeat the same pacification process until he reached the Chesapeake. At the time, Cornwallis's success seemed assured, as he faced resistance only from a handful of poorly armed Patriot partisan detachments, like those led by Francis Marion and Thomas Sumter. Pacifying efforts in the Carolinas quickly went awry as misbehavior by the Tory militia, taking advantage of the British Army presence, further angered the large number of uncommitted Americans hoping to remain neutral during the conflict. Fueled by actual and imagined British excesses, Patriot insurgent attacks grew in frequency throughout the summer of 1780, and fights between Patriot and Tory militia groups assumed the bloody character of a civil war.[6]

Shocked by the Charleston debacle, the Continental Congress reacted by sending General Horatio Gates to organize a new Southern Army near Hillsborough, North Carolina. Gates, "a man of modest talent whose most consistent attribute was an inflated estimation of his own abilities," unwisely chose to march immediately on Camden, instead of taking time to mold a trained and cohesive force of Continentals and militia.[7] Forewarned, Cornwallis quickly regrouped dispersed infantry units from their forward operating bases to reform the field army. Advanced detachments of the two armies bumped into each other in a swampy pine barren during the night of 15–16 August 1780. Surprised by British troops so far from Camden, Gates compounded his error by deploying the tired and dispirited militia as regulars, ordering them into line on the American left, opposite the best British troops. Faced with a disciplined bayonet advance, the poorly equipped Patriot militia folded and fled, allowing the British infantry to enfilade the Continentals while Tarleton's dragoons sliced into them from the rear. In all, the Americans lost around 1,200 casualties killed and wounded, with large numbers of Continentals taken prisoner. Gates shamefully abandoned the army on the battlefield and ran for safety in Charlotte, North Carolina, leaving the dispirited remnants to make their own way back to Hillsborough.[8] With this latest battlefield success, Cornwallis felt the time was almost right to advance his army into North Carolina without facing serious Continental resistance. All that remained was some extra work to mop up the pesky and increasingly better-organized insurgents operating on the flanks and rural areas within the British security zone.[9]

On 19 August, Tarleton's *Legion* won another smashing victory, when Thomas Sumter's partisan command foolishly failed to implement adequate security procedures at their patrol base along Fishing Creek. Tarleton's dragoons easily scattered the partisans, nearly capturing Sumter in the process. On that same day, a mixed battalion of 500 Tory provincial and standing militia was badly beaten by 200 mounted Patriots at Musgrove's Mill, with 63 killed and 70 taken prisoner in exchange for only 4 killed and 12 wounded on the Patriot side. From Cornwallis's perspective, Musgrove's Mill was the latest in the string of battlefield failures by the Loyalist militia, shortfalls brought into stark relief by the superb combat record of British regular units like Tarleton's *Legion*. Despite the rosy promises of the expatriate Loyalists, all indications showed that the Tory militia could not, unsupported, keep the insurgents in check.[10]

---

[6] Dauphinee, "Lord Charles Cornwallis and the Loyalists," 20.

[7] James K. Swisher, *The Revolutionary War in the Southern Backcountry* (Gretna, LA: Pelican Publishing, 2012), 147.

[8] Charles Ross, ed., *Correspondence of Charles, First Marquis*, vol. 1 (London: John Murray, 1859), 59. Contemporary correspondence often used Charlotte-town or Charlotte Town. For continuity's sake, the modern spelling will be used.

[9] Robert W. Brown Jr., *Kings Mountain and Cowpens: Our Victory Was Complete* (Charleston, SC: History Press, 2012), 41.

[10] Dauphinee, "Lord Charles Cornwallis and the Loyalists," 39.

KINGS MOUNTAIN

While Cornwallis had been busy trying to stamp out the insurgency, Major Patrick Ferguson had been busy working to fulfill his mandate to organize the Loyalist militia into an effective internal security force for the British rear areas. Appointed as inspector general of militia by Clinton shortly after the capture of Charleston, Ferguson had been working to organize Loyalists in the Ninety Six district with the help of a battalion of provincials. In the hierarchy of the British Army, provincials were neither regulars nor militia, as they were American-born Loyalists recruited and organized under the authority of the royal governor and enlisted for full-time service only for the duration of the war. Unlike the militia, the provincials were paid, equipped, and drilled to the same standard as regulars and performed just as well during the early battles of the revolution.[11]

Working diligently, Ferguson enrolled around 5,000 soldiers in Loyalist militia regiments; of that number, he enlisted around 1,500 militia of the best, generally single and landless men, into active service for six months. By late summer, Ferguson felt that his troops—drilled in standard British line infantry tactics—were ready for field operations. Ferguson took his troops on patrols into the eastern foothills of the Blue Ridge Mountains to toughen them up and gain some operational experience, incursions which soon aroused the concern of the Patriot Overmountain men of the western Blue Ridge.[12]

Despite his best efforts in training and leadership, Ferguson and his provincial cadre were not able to instill a sense of confidence within the ranks of the Loyalist militia. In 1775–76, better-motivated Patriot militia units easily suppressed Loyalist counterrevolutionary efforts in the Carolinas, instilling a sense of moral inferiority not easily dispelled by a couple months of training.[13] Yet, despite his own reservations and the negative opinions of subordinate commanders, Cornwallis chose to employ the Loyalist militia in an independent offensive role. Cornwallis issued orders to Major Ferguson in late August 1780 to take his Loyalist corps into western North Carolina to recruit new Tory militia companies, protect Loyalists, and in general suppress the rebels. Cornwallis hoped Ferguson's presence in previously unmolested rebel enclaves would overpower the insurgents and facilitate the gathering of the fall harvest into British magazines. By early fall, Cornwallis expected enough improvement to begin the next operational phase of the campaign plan: pacification of North Carolina. At the time a seemingly minor change in the operational plan (so minor that Cornwallis did not clearly notify Clinton of his decision), Cornwallis's redeployment of Ferguson's militia set in motion a chain of events that led to the complete disruption of the British strategy in the Carolinas in the space of six weeks.[14]

Ferguson's corps left Ninety Six in late August 1780, marching 50 miles (80.5 km) northwest along the Broad River to the suspected rebel enclave of Gilbert Town, North Carolina. After setting up camp and recruiting a fresh North Carolina battalion, Ferguson issued a threatening proclamation to the Overmountain rebel settlements—a message that Ferguson would soon have abundant cause to regret: "If they did not desist from their opposition to the British arms . . . he [Ferguson] would march his army over the mountains, hang their leaders, and lay their country [to] waste with fire and sword."[15]

Ferguson's message was delivered to Colonel Isaac Shelby, commander of the Burke County, North Carolina, militia regiment, by Samuel Phillips, a paroled prisoner and kinsman of Shelby's.

[11] Lambert, *South Carolina Loyalists in the American Revolution*, 149–50.

[12] Lambert, *South Carolina Loyalists in the American Revolution*, 138–39.

[13] Howard H. Peckham, ed., *Sources of American Independence: Selected Manuscripts from the Collections of the William H. Clements Library*, vol. 2 (Chicago, IL: University of Chicago Press, 1978), 293.

[14] Lambert, *South Carolina Loyalists in the American Revolution*, 135–36.

[15] J. David Dameron, *King's Mountain: The Defeat of the Loyalists, October 7, 1780* (Cambridge, MA: De Capo Press, 2003), 29–30.

Provoked by Ferguson's threats and presence so close to the frontier settlements, Shelby decided to organize a punitive expedition to eliminate Ferguson and the threat from the Tory militia corps. A group of roughly 1,200 Patriots from North Carolina and Virginia quickly gathered and set out to track down and destroy Ferguson's Tory corps before it could reach safety with Cornwallis. During the next several days, the Patriots and Tory militiamen played a cat-and-mouse game across the Broad River valley of north-central South Carolina. Despite having ample warning of the approaching Patriot force, Ferguson inexplicably dawdled in his withdrawal, seemingly torn between concerns for safety and a desire for battlefield glory. By 6 October, Ferguson's corps was drawn up atop a spur of high ground running from Kings Mountain, awaiting reinforcements from Charlotte and nearby Tory militia units.[16]

After some initial missteps caused by bad intelligence, the Patriot force dispatched a flying column of mounted riflemen in fast pursuit of Ferguson. After a dreary forced march on 7 October 1780 through a driving rain, the Patriot force surrounded Kings Mountain and gained tactical surprise over the Tory force, aided by Ferguson's neglect to ensure suitable defensive preparations.[17] After a bloody hour-long battle marked with bayonet charges and desperate acts of courage on both sides, Major Ferguson was shot dead in the saddle attempting to save his command from utter destruction. The surviving Loyalists were captured by the vengeful Patriots, and many of the hapless Tories were "hurried into oblivion" before the Patriot commanders could restore order in the ranks.[18] In total, the British lost 225 killed, 123 wounded, and 716 captured, with materiel losses including approximately 1,500 rifles and muskets and several wagons with ordnance stores. In return, the Patriots suffered only 28 killed and 62 wounded. The Patriots hastily retreated the next day, fearing retribution by Tarleton's dragoons, leaving the Tory wounded on the hill and the dead hastily buried in a shallow draw.[19] The prisoners were forced to carry away the valuable stock of captured firearms, after their flints had been carefully removed.[20]

THE COWPENS

Stunned by the news of the Kings Mountain disaster, and with Tarleton seriously ill, Cornwallis could make no meaningful effort to pursue the rebels, and the entire Patriot force reached safety across the flooded Catawba River early on 16 October. Tired and fearful of pursuit, the Patriots allowed many of their prisoners to escape while on the march. Only a small remnant of 130 prisoners was delivered to the Continental Army at Hillsborough, North Carolina, by a contingent of Virginia riflemen.[21]

Overextended, short of supplies, suddenly bereft of flank security, and plagued by sickness in the ranks, Cornwallis had little choice but to suspend offensive plans and to withdraw his field army to winter quarters in Winnsboro, South Carolina.[22] News of Ferguson's destruction further depressed Tory morale, and recruiting dropped off as fewer Loyalists were willing to risk their lives and property openly in service to the king. Once Cornwallis arrived back at Winnsboro, he assumed the Americans would also suspend their offensive operation to rest and refit for better campaigning weather in the spring of 1781. The new Continental commander of the Southern Army, Major General Nathanael Greene, was not so obliging. Greene had been nominated to replace Major General Horatio Gates

[16] Peckham, *Sources of American Independence*, 295.
[17] Brown, *Kings Mountain and Cowpens*, 60.
[18] Wilma Dykeman, *The Battle of Kings Mountain, 1780: With Fire and Sword* (Washington, DC: National Park Service, U.S. Department of the Interior, 1991), 65–66.
[19] A *draw* is the low ground between two ridges or spurs, not formed by erosion.
[20] Dameron, *King's Mountain*, 82–83.
[21] Robert M. Dunkerley, *Kings Mountain Walking Tour* Guide (Pittsburgh, PA: Dorrance Publishing, 2003), 26–27.
[22] Dykeman, *The Battle of Kings Mountain, 1780*, 60–74.

in the aftermath of the Camden calamity, but he had not arrived to take charge officially of the Southern Department until 3 December 1780. Greene's first act as commander was to conduct a quick assessment of the strategic and operational situation, which included a thorough engineering and quartermaster assessment of the theater of operations. Realizing his force was too week to directly challenge Cornwallis, Greene divided his forces, partially to reduce the supply burden, but most importantly to seize the initiative from the enemy. Brigadier General Daniel Morgan crossed into South Carolina in late December 1780 at the head of the flying army, Continental infantry and dragoons supported by light infantry militia. The strategically astute Greene understood that the risks of dividing his army were outweighed by the potential gains: "This force . . . you will employ against the enemy on the West side of the Catawba River. . . . The object of this detachment is to give protection to that part of the country and spirit up the people—to annoy the enemy . . . to collect the provision and forage out of their way. . . . [you] will fall upon the flank, or into the rear of the enemy, as occasion may require."[23] In another letter, Greene outlined his strategic premise, noting that Cornwallis could not easily attack either American force without leaving Ninety Six and other bases vulnerable to attack.

> I am well satisfied with the movement [division of the army], for it has answered thus far all the purposes from which I intended it. It makes the most of my inferior force, for it compels my adversary to divide his, and holds him in doubt as to his own line of conduct. He cannot leave Morgan behind him to come at me, or his posts of Ninety Six and Augusta would be exposed. And he cannot chase Morgan far, or prosecute his views upon Virginia, while I am here with the whole country open before me. I am as near to Charleston as he is, and as near to Hillsborough as I was at Charlotte; so that I am in no danger of my being cut off from my reinforcements, while an uncertainty as to my future designs has made it necessary to leave a large detachment of the enemy's late reinforcements in Charleston, and move the rest up on this side of the Wateree. But although there is nothing to obstruct my march to Charleston, I am far from having such a design in contemplation in the present relative positions and strength of the two armies.[24]

Aroused by Greene's audacious plan, Cornwallis detached the reinforced *British Legion* to pursue and destroy Morgan's detachment, while ordering up new reinforcements from Charleston. Starting in early January 1781, Tarleton rode his command hard, first to screen Ninety Six, before turning to pursue Morgan. Morgan learned of Tarleton's pursuit and began a hasty withdrawal toward the Broad River and safety. Tarleton relentlessly pushed his troops without hot food or rest and by the afternoon of 16 January, Morgan realized he would not reach sanctuary in time. He chose to fight, diverting his command to a suitable parcel of land for defense, the Cowpens. Morgan was a skilled tactical commander, and he used the evening of 16 January to rest his force, while Morgan himself went from camp to camp to explain to his soldiers the plan of battle. At daybreak on 17 January 1781, Tarleton's command deployed into an immediate attack against the Americans. Expecting an aggressive frontal attack by Tarleton, Morgan cannily deployed his rifle-armed militia in a way that maximized their skirmishing strengths without exposing them to needless risk from British bayonets. Shaken by the unexpectedly heavy militia fire, the cohesion of the tired and hungry British infantry was shattered by Continental volley fires and enveloped by the regrouped militia riflemen and Con-

[23] William Johnson, *Sketches of the Life and Correspondence of Nathanael Greene, Major General of the Armies of the United States in the War of the Revolution*, vol. 1 (Charleston, SC: A. E. Miller, 1822), 346–47.
[24] Lawrence E. Babits and Joshua B. Howard, *Long, Obstinate, and Bloody: The Battle of Guilford Courthouse* (Chapel Hill: University of North Carolina Press, 2009), 9–10.

tinental dragoons. Tarleton was unable to rally his troops and fled the Cowpens with a remnant of *Legion* dragoons, leaving behind some 300 dead and 525 prisoners.

THE RACE FOR THE DAN

Elated by his victory, Morgan policed up the prisoners and moveable wounded and fled for sanctuary across the Broad River. Even as Cornwallis absorbed the shock of Tarleton's defeat, Major General Sir Alexander Leslie arrived from Charleston with reinforcements: two elite *Guards* battalions, the *Von Bose Regiment*, and a detachment of German jägers—approximately 1,500 fresh fighting troops. Thus reinforced, Cornwallis set out to destroy Morgan's corps and recover the *Legion* prisoners. The relative lack of mounted troops and Tory support bedeviled Cornwallis in his pursuit of the Americans. After a week of hard marching in the cold and wet, Cornwallis decided to burn his excess wagons and baggage, sparing only enough wagons for salt, ordnance stores, and wounded. Morgan took advantage of Cornwallis's slow pursuit to cross the Catawba River and link up with Greene. Faced with Cornwallis's superior army, Greene had little choice but to withdraw toward sanctuary in Virginia, the slower-moving main body and logistics trains screened by detachments of militia and light infantry. All the while, Greene kept his militia detachments busy sweeping the region clear of supplies in a bid to stretch Cornwallis's army to its culmination point. Once Greene reached sanctuary in Virginia, he then hoped to receive enough Continental and militia reinforcements to attack the suitably weakened British.

After a short pause to gather supplies and prepare for the next operational phase, Cornwallis came out swinging. His soldiers forced a crossing of the swollen Catawba River on 1 February, easily brushing aside Whig militiamen on the riverbanks and during a skirmish at nearby Tarrant's Tavern. Beset by the weather, both armies force-marched toward the Yadkin River in the beginning stages of the Race to the Dan. Greene's close attention to logistics and intelligence made the difference between safety and victory. For example, Greene realized in his first reconnaissance of the Carolinas that rivers and creeks posed a great impediment to movement. Accordingly, Greene had a detail of engineering troops construct a flotilla of bateaux (light flat-bottomed boats) for moving his troops across flooded rivers. The train of bateaux accompanied the Continental force and gave the Americans mobility unmatched by the British. Besides lacking engineering assets, Cornwallis was handicapped by poor intelligence about Greene's movements and engineering reconnaissance data about suitable river crossing points. Despite some close calls, the American Army successfully beat the British Army across the Dan River on the night of 14–15 February 1781. Greene had skillfully lured the headstrong Cornwallis into a corner, leaving the British Army hungry, tired, and days from the nearest supply depot. By contrast, Greene was in friendly country, close to supply depots and militia reinforcements from Virginia.[25]

HARE AND HOUNDS WITH CORNWALLIS

Having little choice in the matter, Cornwallis withdrew his army toward the Tory enclave of Hillsborough, North Carolina, to rest and gather supplies. Greene was determined not to give Cornwallis time to recover by sending a provisional light corps, made of Lieutenant Colonel Henry Lee's Legion and a militia brigade commanded by Brigadier General Andrew Pickens, back into North Carolina in mid-February 1781. Greene intended to employ his light forces to further weaken Cornwallis by

---

[25] Babits and Howard, *Long, Obstinate, and Bloody*, 36.

shutting off supplies and Tory support within the region. Within days, a 400-man battalion of Tories under Colonel John Pyle met Lee's troopers near the Haw River. The inexperienced Tory troops initially mistook Lee's green-jacketed fighters for Tarleton's *Legion*, not learning of their mistake until too late. The ensuing carnage—some 90 Tories killed, and the remainder dispersed—was later dubbed Pyle's Hacking Match. This latest Loyalist disaster took place within miles of Tarleton's patrol base, laying bare the inability of the British to protect their Loyalist supporters. What little support remained from organized Tory militia evaporated, and Cornwallis was bereft of friendly militia support for the remainder of the campaign.[26]

Cornwallis soon moved from Hillsborough to the Alamance Creek area to find fresh supplies and to keep his troops from "straggling out of Camp in search of whiskey."[27] Encouraged by Cornwallis's troubles, Greene moved his main army across the Dan River on 22 February to begin a cat-and-mouse game with the British as additional American militia reinforcements kept arriving. Light forces on both sides were kept busy screening and skirmishing, while Cornwallis strove to find the American main body: "From that time to the 15th of March the two armys were never above twenty miles [32.2 km] asunder, they [were] constantly avoiding a general action and we as industriously seeking it. These operations obliged the two armys to make numberless moves which it is impossible to detail."[28] The constant tension and cold weather stressed both armies. The British fighters, reduced to eating turnips or parched corn, felt the pinch of genuine hunger, while resentment at the absence of rum caused increased straggling. On the American side, Greene faced problems with the Carolina militia, who resented their treatment by Continental officers. Matters came to a head when Cornwallis's attempt to trap and destroy the American light forces resulted in a sharp skirmish at Weitzel's Mill on 6 March 1781. Although the Americans avoided major losses, South Carolina and Georgia militia felt they had been left unsupported on the battlefield while covering the withdrawal of the Continental line. As a result, Pickens and most of the South Carolina militia left the army for home, leaving Greene dependent on North Carolina and Virginia militia of unknown quality.

OPENING DISPOSITIONS

Despite his challenges, Greene felt he was strong enough with the inclusion of two North Carolina militia brigades and several Virginia state regiments to offer battle in early March. After quick preparations, Greene marched his consolidated army to Guilford Courthouse on 13 March 1781, a location he had previously surveyed and selected as a suitable site to fight Cornwallis, deployed on rough and wooded terrain to the west of the courthouse overlooking the New Garden Road. Cornwallis soon learned of Greene's location and decided to move to contact on the morning of 15 March 1781. Lee's corps of observation, the Partisan Legion with infantry attachments, screened the main American force, and the light troops fought a series of skirmishes along the New Garden Road about 4 miles (6.4 km) to the west of Guilford Courthouse. Amply warned by Lee, Greene had arranged his force of roughly 4,400 troops in three lines. The first, made of two North Carolina militia brigades, flanked by companies of Continentals and Virginia riflemen, faced open fields straddling the New Market Road. The second line of Virginia state troops was deployed within a belt of timber, and the final Continental line overlooked an irregular farm field close to the actual courthouse.

---

[26] Babits and Howard, *Long, Obstinate, and Bloody*, 38–39.
[27] Babits and Howard, *Long, Obstinate, and Bloody*, 42.
[28] From a letter written by BGen Charles O'Hara after the battle. Buchanan, *The Road to Guilford Courthouse*, 365.

THE FIRST LINE ENGAGEMENT

Beginning around 1330 that afternoon, Cornwallis's army of 1,950 regular soldiers crossed Horsepen Creek and deployed into line formation between the creek and the Hoskins farmstead, covered by the harassing fire of 3-pounder cannons. At the first line, the British were hit by at least one volley from North Carolina militia, supported by aimed rifle fire from the flanking companies. After absorbing the volley, the British infantry charged and broke through the North Carolina militia, who lacked the bayonets and training sufficient to withstand such an attack. Even as the Carolina militia stampeded to safety, enfilade fire from intact American units on the flanks of the breakthrough caused the British line to split into two distinct parts. The bulk of the British heavy and light infantry pushed to the east into the belt of woods to engage the Virginia regiments, while the *Von Bose* and *1st Battalion, Brigade of Guards*, angled off on a diverging axis in pursuit of Lee's Legion and its supporting Virginia riflemen.

THE SECOND LINE ENGAGEMENT

The British *23d* and *33d Regiments of Foot*, supported by British light infantry and jägers, made a disjointed approach toward the Virginia Line. Lawson, the Virginia brigade commander to the north, ordered his regiments to flank the supposedly open left flank of the *23d Foot*. As the Virginians moved forward, they were in turn flanked by the *23d Foot* and supporting *2d Battalion, Brigade of Guards*, battalion. British in the rear of the Virginia regiments caused such confusion that portions of both regiments "instantly broke off without firing a single gun, and dispersed like a flock of sheep frightened by dogs."[29] Virginia officers attempted to rally their troops, and a disorganized mix of platoons and companies fired close-range volleys with the equally disorganized *23d* and *33d Foot*. Despite the disarray in their ranks, the British prevailed due to their bayonets and better training and quickly pushed the mix of Virginians and Continentals to the north and east.

Meanwhile, Brigadier General Edward Stevens's Virginia brigade to the south of the road fought the *2d Guards* battalion and elements of the *71st Foot (Highlanders)* and *Von Bose Regiments*. Large unit cohesion also broke down in the thick woods, and the engagement south of the road dissolved into a confused free-for-all of small unit actions fought at close range. Compared to the other militia brigades, Stevens's brigade fought well and significantly delayed the British before withdrawing in disorder before the British bayonets.

THE BRITISH ENGAGE THE CONTINENTAL LINE

After a confused melee in the woods, the disorganized British left flank began to emerge in the open area to the west of the Continental third line around 1630. Here, Greene had his Continentals deployed to take advantage of the available cover and concealment. The inexperienced 2d Maryland Regiment faced southwest into the angle of the Great Salisbury and Reedy Fork Roads. The battle-tested 1st Maryland Regiment faced almost due west overlooking the open fields. Brigadier General Isaac Huger's Virginia Continental Brigade, consisting of Lieutenant Colonel Samuel Hawes's 2d Virginia Regiment and Colonel John Green's 1st Virginia Regiment, arced to the northeast, screening the Reedy Fork Road. Anchoring the flanks was Lieutenant Colonel William Washington's Corps of Observation and a handful of intact Continental and Virginia militia companies that had withdrawn from the second line. Overlooking the expected enemy avenue of approach was a Continental artillery section.

---

[29] George Washington Coleman Jr., ed., "The Southern Campaign, 1781, from Guilford Court House to the Siege of York, Narrated in the Letters by Judge St. George Tucker to His Wife," in *The Magazine of American History*, vol. 7 (New York: A. S. Barnes, 1881), 40.

The *33d Foot*, led by Lieutenant Colonel James Webster, supported only by the jäger and *Guards* light infantry companies, attacked toward the only visible Continentals, the 2d Maryland. The *33d Foot* was immediately driven back with severe casualties by enfilade fire from the veteran 1st Maryland and grapeshot from Continental 6-pounder cannon fire. Wounded, Webster withdrew his battered regiment to the west edge of the open swale to assume a hasty support-by-fire position. Greene attempted to launch a counterattack, but the green 2d Maryland soldiers "were repeatedly led up to action by Genl. Greene in person, and could not be induced to stand the fire of the enemy: the presence of the general seconded by the exertions of Huger, Hawes and Campbell."[30]

Around the same time that the Continentals repelled the *33d Foot*, Brigadier General Charles O'Hara ordered the *2d Guards* battalion, moving parallel to the Salisbury Road, to attack toward the 2d Maryland and the visible Continental artillery section. The *2d Guards* battalion's attack hit the 2d Maryland in the right flank "when, unaccountably, the second regiment gave way, abandoning to the enemy the two field-pieces."[31] In overrunning the Marylanders, the *Guards* were in turn hit by the hidden 1st Maryland, which had faced about to reverse its battle line. As the *Guards* and 1st Maryland regiments, arguably the best regiments in each army, traded volleys in the open vale, Lieutenant Colonel Washington led his dragoon squadron into a slashing attack into the rear of the *Guards*. Close-range fighting resulted in heavy casualties, and many officers fell dead or wounded.

Cornwallis arrived on the west edge of the swale around the time the shattered *Guards* regiment was also withdrawing toward Webster's *33d Foot*. Seeing the danger from Washington's dragoons, Cornwallis ordered his accompanying 3-pounder guns to fire grapeshot to drive off the Continental dragoons and infantry. As the Americans withdrew back toward the courthouse, more British units—the *23d* and *71st Regiments of Foot*—reached the scene, and Cornwallis arrayed them in attack formation. By this point, Greene had ordered the withdrawal of the Continental line after witnessing the folding of the 2d Maryland. The unengaged 1st Virginia was designated as the delaying force across the Reedy Fork Road. Among the first Continental units to withdraw was the 2d Virginia and the disorganized elements of the 2d Maryland. As the Continentals withdrew, Greene had no choice but to abandon his artillery, as many of the horses "had been mostly killed" by enemy fire.[32] Believing victory was at hand, Cornwallis ordered the *23d* and *71st Foot* and the *2d Guards* battalion forward in pursuit. A couple of disciplined volleys from the 1st Virginia and 1st Maryland kept the British at bay, and Cornwallis soon broke off the pursuit due to the exhausted state of his troops.

THE FIGHT IN THE SOUTH

While the main fight at the second and third line raged, the *1st Guards* battalion and *Von Bose Regiments* fought a separate action with Lee's Corps of Observation, strengthened by the addition of Major Alexander Stuart's Virginia militia battalion that had withdrawn intact from the wrecking of Stevens's militia brigade. The *1st Guards* battalion suffered substantial casualties as Stuart's Virginians skillfully fought a delaying action using a reverse slope position. Many British officers fell dead or wounded, and command and control began to collapse in the *Guards*. At this juncture, the *Von Bose* outflanked the Virginia militia, breaking their lines. As the *1st Guards* and *Von Bose* reformed and drove southeast, fire from American units hit their flanks and even the rear. In response, the *Von Bose* commander, Major du Buy, wheeled two companies "to attack towards rear, against the enemy, who were behind

[30] Babits and Howard, *Long, Obstinate, and Bloody*, 145.
[31] Henry Lee, *Memoirs of the War in the Southern Department of the United States* (New York: University Publishing, 1869), 280.
[32] Richard K. Showman, Dennis M. Conrad, and Roger N. Parks, eds., *The Papers of General Nathanael Greene*, vol. 7 (Chapel Hill: University of North Carolina Press, 1994), 435.

us, and forced them again to take flight."[33] Smoldering cartridge papers set some brush afire, reducing visibility further. After about 30 minutes of bloody close-range fighting, the "battle within a battle" had veered far to the south, perpendicular to the Great Salisbury Road, and continued even as Greene's Continental line broke contact.[34] Alerted by the continued firing to the south, Cornwallis ordered Tarleton to take a dragoon squadron to reconnoiter. Tarleton probably returned to the trace of the first line fight, before riding south and east. Tarleton arrived on the scene of battle around the time that Lieutenant Colonel Lee decided to withdraw toward the courthouse, a questionable decision that left Campbell's Virginia riflemen alone and unsupported on the battlefield. The *1st Guards* battalion and *Von Bose Regiment* fired a volley at the largest group of Americans, and Tarleton charged on the right flank. Left unsupported by Lee's dragoons, the Virginia infantrymen had no choice but to run for their lives. Many Americans were ridden down and sabered, although some were able to shoot back, and one scored a lucky hit by shooting two fingers off Tarleton's right hand. After a short pursuit, Tarleton returned his dragoons to Cornwallis's position near the courthouse. Left unmolested, Lee's Legion avoided enemy contact and rejoined Greene's main body somewhere north of the courthouse.[35]

THE AFTERMATH

Although orderly, the American withdrawal left the British in possession of the field of battle and several Continental artillery pieces. Greene's army withdrew to a designated staging point at Speedwell Ironworks, 13 miles (21 km) from Guilford Courthouse. Despite the collapse of several American units on the battlefield, officers quickly restored order, and the movement was organized and done without incident. Stragglers were policed up, and as units arrived in camp, the soldiers threw up hasty fortifications, followed by the issuing of two days' rations plus a gill of rum.[36] Although tactically victorious, Cornwallis's command was too badly damaged for any type of pursuit and exploitation operation. Instead, the tired and hungry British camped on the battlefield under a driving rainstorm that prevented the building of fires. The rain continued to fall unabated through the next day, and the wet and exposure compounded the misery of the wounded. Cornwallis was burdened with several hundred wounded from both sides. Out of 1,924 British and Hessians on the battlefield, Cornwallis lost 93 killed, 413 wounded, and 26 missing, a 28-percent casualty rate. Although Greene's manpower returns are suspect, he reported 79 dead, 184 wounded, and 1,046 missing out of approximately 4,400 soldiers; of the Continentals: 57 killed, 111 wounded, and 161 missing.[37]

Redcoats, aided by local Quaker citizens, worked through the night and next day to evacuate the transportable wounded to nearby houses. Under a flag of truce, Greene supplied surgeons and food to help minister to the hundreds of wounded collected in houses at New Garden Meeting House and Guilford Courthouse.[38] After a couple days of treating the wounded and trying to find supplies, Cornwallis realized that the promised Loyalists would never appear to help his beleaguered army. Bereft of supplies, isolated in a hostile country, and lacking the essential support of the Tory militia, Cornwallis had little choice after his Pyrrhic victory at Guilford Courthouse but to abandon "his

---

[33] Babits and Howard, *Long, Obstinate, and Bloody*, 134.
[34] Babits and Howard, *Long, Obstinate, and Bloody*, 129.
[35] John R. Maass, *The Battle of Guilford Courthouse: A Most Desperate Engagement* (Charleston, SC: History Press, 2020), 151.
[36] A *gill* as a unit of measurement was equivalent to about one-quarter pint, about a teacup's worth.
[37] Babits and Howard, *Long, Obstinate, and Bloody*, 173, 175. The authors note that Greene's official returns missed many North Carolina and Virginia militia casualties and estimate the actual casualties were 94 dead and 220 wounded.
[38] Babits and Howard, *Long, Obstinate, and Bloody*, 179.

mad scheme" to conquer the Carolinas and withdraw to the port of Wilmington to rest and refit.[39] After refitting his shattered army, Cornwallis marched it north, abandoning the Carolinas to seek a decisive victory against the Continentals somewhere in Virginia—a campaign that ended in ignominy at Yorktown in October 1781.[40]

Cornwallis's departure left the British garrisons in South Carolina isolated and unsupported. Greene gladly seized the initiative forfeited by Cornwallis and marched his Continentals into South Carolina, where he worked cooperatively with the Patriot militia bands to eliminate the British forward operating bases. Despite some tactical reverses (Greene never won a single battle while in command of the Southern Army), the lack of an organized militia and sufficient field forces left the British no good options in the Carolinas. By April 1781, the British had entirely ceded control of the interior, holding on only to the coastal enclave of Charleston. The last major battle at Eutaw Springs in South Carolina in September 1781 cemented American control of the countryside, and aside from a few insignificant Tory raids, the British remained isolated and impotent in Charleston until the end of the war.[41]

[39] Showman, Conrad, and Parks, *The Papers of General Nathanael Greene*, 435. In a letter to Gen Isaac Huger dated 30 January 1781 (after the Cowpens battle), Greene wrote: "It is necessary we should take every possible precaution to guard against a misfortune. But I am not without hopes of ruining Lord Cornwallis if he persists in his mad scheme in pushing through the Country, and it is my earnest desire to form a junction as soon as possible for this purpose." *The Papers of General Nathanael Greene*, 220.
[40] Buchanan, *The Road to Guilford Courthouse*, 383.
[41] Richard W. Stewart, ed., *American Military History*, vol. 1, *The United States Army and the Forging of a Nation, 1775–1917*, 2d ed. (Washington, DC: U.S. Army Center of Military History, 2009), 96–99.

PART III

FIELD STUDY PHASE:
SUGGESTED ROUTES AND STANDS

INTRODUCTION

The second—field—phase of the staff ride consists of an instructor-led discussion of the Guilford Courthouse campaign, with each stop (or *stand*) conducted on or overlooking the terrain where the action took place. The field study phase is intended to help the student better understand the historical events before, during, and after the battle; critically analyze the significance of those events; and, from that analysis, distill relevant insights into the profession of arms. As no campaign takes place in isolation, the instructor should devote considerable attention to explaining individual actions within the proper strategic and operational context. The Guilford Courthouse campaign stretched across a roughly 200-mile (322-km) swath of territory from up-county South Carolina to the Virginia-North Carolina border; thus, this guide is structured to examine the major points of the campaign in a two-day field study phase. The suggested itinerary for the first day starts at the Cowpens National Battlefield and ends with the dragoon skirmishes before the start of the battle at Guilford Courthouse. With 30 minutes allotted for each stand, the first-day staff ride should take about 10 hours to complete, not counting meal breaks. The second day itinerary focuses on the tactical actions during and after the battle, with all stands conducted at Guilford Courthouse National Military Park (GCNMP). The day two itinerary should cover about nine hours, not counting meal breaks. For those units constrained by time and resources, the handbook also suggests shorter itineraries. The operational-level stands outside of GCNMP are generally in publicly accessible areas, with the exceptions explained later in the text.

GUILFORD COURTHOUSE
CAMPAIGN STAFF RIDE STANDS

Day 1

Stand 1–Campaign overview at Cowpens National Battlefield

Stand 2–Cowan's Ford

Stand 3–Trading Ford

Stand 4–Boyd's Ferry

Stand 5–Whitesell's Mill

Stand 6–New Garden Meeting House

Day 2

Stand 1–Operational overview at GCNMP

Stand 2–Tactical situation

Stand 3–Hoskins Farmstead

Stand 4–The American first line

Stand 5–Breaking of the first line

Stand 6–The fragmented attack

Stand 7–Lawson's brigade

Stand 8–Stevens's brigade

Stand 9–The fight in the South

Stand 10–The third line fight

Stand 11–The fight for the cannons, British perspective

Stand 12–The fight for the cannons, American perspective

Stand 13–The Continental withdrawal

Stand 14–The butcher's bill

MODIFICATIONS TO THE TWO-DAY STAFF RIDE

Although the Guilford Courthouse campaign is best studied during two days, military professionals can still gain much of value with a shorter staff ride. The easiest way to trim the campaign-level staff ride is to eliminate those stands that are not within close distance to GCNMP. One option is to bypass the operational-level stands to focus mostly on the tactical-level actions—an approach best suited for a training audience made of junior leaders or precommission cadets or candidates. Another good option for tailoring a shorter event is to eliminate selected stands outside of GCNMP to focus on specific teaching points, such as the sustainment or protection warfighting functions. Option one below is more focused on operational-level decision-making by Cornwallis and Greene, while option two, which includes the two skirmishes before the Guilford Courthouse battle, is best suited for a tactically focused staff ride.

Option 1: Operational Focus One-Day Guilford Courthouse Campaign Staff Ride

Stand 1–Campaign overview at Cowpens National Battlefield

Stand 2–Cowan's Ford

Stand 3–Trading Ford

Stand 4–Hoskins farmstead

Stand 5–The American first line

Stand 6–Breaking of the first line

Stand 7–The fragmented attack

Stand 8–Lawson's brigade

Stand 9–Stevens's brigade

Stand 10–The fight in the South

Stand 11–The third line fight

Stand 12–The fight for the cannons, British perspective

Stand 13–The fight for the cannons, American perspective

Stand 14–The Continental withdrawal

Stand 15–The butcher's bill

Option 2: Tactical Focus One-Day Guilford Courthouse Campaign Staff Ride

Stand 1–Campaign overview at GCNMP

Stand 2–Whitesell's Mill

Stand 3–New Garden Meeting House

Stand 4–Hoskins farmstead

Stand 5–The American first line

Stand 6–Breaking of the first line

Stand 7–The fragmented attack

Stand 8–Lawson's brigade

Stand 9–Stevens's brigade

Stand 10–The fight in the South

Stand 11–The third line fight

Stand 12–The fight for the cannons, British perspective

Stand 13–The fight for the cannons, American perspective

Stand 14–The Continental withdrawal

Stand 15–The butcher's bill

SEQUENCING OF STANDS

For the two-day staff ride event, the first set of operational stands are sequenced chronologically, beginning just after the Battle of the Cowpens in January 1781, and ending with the New Garden Meeting House skirmish early on 15 March. The suggested route for the second day's tactically focused Guilford Courthouse battlefield staff ride incorporates 14 stands that are all accessible on foot within the bounds of the GCNMP. A suggested route map is included that provides a logical flow of stands with minimal backtracking. The first stand on each day incorporates a visit to the respective national park visitor's center, conducive to a quick refresher of the strategic and operational context for the campaign.

Each stand is scripted in a standardized format that will lend an orderly flow to the terrain walk. First, directions and a map help the instructor guide students from one stand location to the next. The orientation paragraph aids the instructor in accurately depicting the location of the historical actions in relation to the adjacent terrain. The description section provides a narrative guide to historical actions during the battle, which are tied to the terrain in the vicinity of the stand. Following the description are one or more primary-source vignettes, human interest stories that when read aloud illuminate the face of battle and give insights into the timeless human factors present in combat. Last, suggested analysis questions help the instructor stimulate critical thinking and discussion about the *how* and *why* of the events discussed at the stand. All azimuths given in the text are magnetic, while times used in the text are given in a 24-hour format, and local to the Guilford Courthouse region (U.S. eastern time zone).

STAFF RIDE METHODOLOGY

The field study phase consists of walking to particular locations on the battlefield, known as stands, and participating in a discussion of events using the orientation, description, and analysis (ODA) structuring technique discussed earlier. Sequencing of the stands is done so that by the end of the terrain walk the students have had time to thoroughly analyze all aspects of the battle.

Orientation

The purpose of the orientation is to ensure that participants understand the physical characteristics of the location as it was during the battle. The instructor should ensure that the initial orientation at the start of the terrain walk is detailed, with subsequent orientation tailored to the needs of the students. Suggested minimums to include at each stand include:

A. Location of the previous stand. If visible, point it out. If beyond line of sight, use a map board and cardinal directions.
B. The current location on the map, with key terrain features.
C. Weather and light data, time of day, and season at the time of the battle.
D. Approximate location of units in relation to the terrain and each other.
E. Permanent or temporary structures that were present during the battle.
F. Significant changes in terrain, vegetation, or structures from 1781 to present.

During the terrain walk, the instructor can gauge student situational awareness by asking simple open-ended questions. For example, "Which way is north?"

Description

The purpose of the description paragraph is to ensure students are familiar with the historical events that occurred in the vicinity of the stand location. A suggested technique is to use a chronological review of the following key elements:

A. Unit movements.

B. Combat actions: attacks, maneuvers, defends, and withdraws.

C. Leader movements, location, decisions, and action.

D. Individual soldier acts of bravery or cowardice.

The instructor should minimize lecturing, and as much as possible allow students to provide their own descriptions, based on their preliminary studies, as a means of maximizing experiential learning. One such method is to assign a specific study area to each student: a historical figure (e.g., Lieutenant General Lord Charles Cornwallis or Lieutenant Colonel Henry Lee); warfighting functions (e.g., movement and maneuver, fires, intelligence, protection, sustainment, and mission command/leadership); or specific military branch functions (e.g., cavalry, infantry, or quartermaster).

Analysis

Critical analysis at each stand is the central point of the terrain walk. Open-ended questions are directed at the students to get them to consider the circumstances behind a particular leader action. Ideally, critical analysis will help the student gain some insight into the timeless aspects of combat leadership. Here, the instructor should strive to stimulate critical thinking skills in the students, such as asking for analysis of character actions or their findings based on their specific study area. Role-playing here is particularly good at drawing out insights, as the role player should try to explain the character's historical decision-making, based on the information at hand. During this portion, a brainstorming group approach is useful, with errors in fact or logic tactfully corrected by the instructor. Generally speaking, analytical questions fall into two categories: historical evaluations and modern relevancies.

1. Historical evaluations. Here, the instructor guides the students through questions regarding the historical leaders, their units, and systems in their historical context. More than "Monday-morning quarterbacking," questions here should focus on the facts, assumptions, and mental and physical factors that influenced a leader's decision. In particular, analysis should center on the exercise of command and control, how leaders trained and prepared for combat, and how weapon systems were integrated into combat.

2. Modern relevancies. Here, the instructor helps participants draw out relevant insights for the military professional. One method is to view historical factors within the context of modern doctrine. For example: consider hasty versus deliberate mission planning in the context of the historical attack plan. What could they have done differently? Or, using *Offense and Defense*, Army Doctrine Reference Publication (ADRP) 3-90, chapter four, or *Commander's Tactical Handbook*, Marine Corps Reference Publication (MCRP) 3-30.7 (formerly MCRP 3-11.1A), "Defensive Operations" chapter, as a guide, what were some defensive planning shortfalls? Open-ended questions here are particularly useful, such as: Why did a particular action fail or succeed? or, What were some of the factors leading to the successful friendly attack? Avoid asking simple questions answered with a yes or no.

Encourage debate and disagreement among students, and certainly point out that there are often multiple points of view to an issue. The instructor should be prepared to respectfully challenge students to logically defend their arguments. Here, humility and tact are essential, as at times even a junior student can reveal fresh insights not readily apparent even to an experienced instructor. Also, the instructor should encourage students to carry a notebook to record their thoughts and observations *as they occur*, rather than waiting for the end of the terrain walk, which will make for a more productive integration session.[1]

DAY ONE: GUILFORD COURTHOUSE CAMPAIGN STAFF RIDE STANDS

Stand 1: Campaign Overview

Introduction

The inclusion of operational-level stands allows a unit commander to tailor the Guilford Courthouse staff ride toward a campaign-level leader development event. To ensure the best setting for the event, the facilitator must ensure that a thorough strategic and operational refresher is done before starting the stands. A suggested start point for the operational orientation is to begin at the Cowpens National Battlefield located at 4001 Chesnee Highway, Gaffney, SC 29341. The victory of Brigadier General Daniel Morgan's light corps of 1,100 troops over the 900 soldiers of Lieutenant Colonel Banastre Tarleton's *British Legion* on 17 January 1781 set in motion the train of events that was to end at Guilford Courthouse on 15 March 1781.

Orientation

As this is the beginning of a campaign-level overview, the use of good-quality maps and supporting graphics are essential in helping students understand the area of operations. On the map, the instructor should point out the major British bases at Charleston, Ninety Six, and Camden. On 17 January 1781, Cornwallis's main army was camped about 45 miles (72 km) to the southeast, adjacent to Turkey Creek. The light forces of both armies, Brigadier General Morgan's flying army and Lieutenant Colonel Banastre Tarleton's *British Legion,* met in battle at the Cowpens at daylight on 17 January. The Americans' primary forward logistics depot was at Salisbury, North Carolina, about 100 miles (161 km) to the northeast, while General Nathanael Greene's army of Continental regulars was camped at Cheraw Hill, about 150 miles (241 km) to the east.

Description

Before beginning the staff ride, it is important to briefly summarize the operational situation for both sides. By late 1779, the surrender of British general John Burgoyne's army at Saratoga had encouraged France and its allies to enter the war against Britain. Left with few good strategic options, Lieutenant General Henry Clinton was prodded by Prime Minister North into launching a risky offensive to retake the southern states. Despite some stunning successes, notably the seizure of Charleston, South Carolina, the British struggled to effectively consolidate their gains. First, Clinton lacked sufficient power to secure all of British-occupied America, so Cornwallis was expected to organize reliable Loyalist militia units to free up regular units for offensive employment. In August 1780, Cornwallis upset an American counteroffensive by destroying General Horatio Gates's Southern Army at Camden. Afterward, Cornwallis abandoned his key task of pacification in favor of an offensive into North Carolina. Cornwallis's plans were upset when Major Patrick Ferguson's first-rate Loyalist militia was destroyed by Patriot militia at Kings Mountain in October 1780. Alarmed by the sudden loss, Cornwallis withdrew his fever-stricken corps to winter quarters back in South Carolina.

While Cornwallis's troops were settling into winter quarters, General Nathanael Greene quietly relieved Gates as commander of the Continental Southern Army. Greene thoroughly assessed the

<hr>

[1] Much of the information for this section is summarized from Charles D. Collins Jr., *Staff Ride Handbook and Atlas: Battle of White Bird Canyon, 17 June 1877* (Fort Leavenworth, KS: U.S. Army Combat Studies Institute Press, 2014), 8–13.

physical and human terrain within the department. Besides this personal reconnaissance, Greene dispatched officers to assess lines of communication and the loyalty of citizens in each district, while making advance arrangements for procurement and delivery of commissary, ordnance, and quarter-master stores. His assessment complete, Greene realized he faced two major challenges: first, the lack of supplies in the region, and second, the American force's weakness in relation to Cornwallis's army. But Greene realized he could not simply focus on gathering supplies and recruits; he had to somehow attain the strategic initiative in the region to prevent the British from consolidating their hold on South Carolina and Georgia. In response, Greene drew up an unorthodox plan to regain the strategic initiative in the Carolinas. He gave Brigadier General Morgan, arguably one of the best light infantry commanders in the Continental Army, command of the flying army of light infantry and dragoons. Greene instructed Morgan to operate west of the Catawba River in upcountry South Carolina, de-nying provisions to the British, encouraging Patriot sympathizers, and threatening the major British logistics base at Ninety Six. After Morgan's departure from Hillsborough, North Carolina, Greene moved the remainder of the army southeast to Cheraw Hill, South Carolina, to threaten British communications between Charleston and Camden. Greene's decision to split his army was a brilliant upending of the principle of mass. The divided elements were better able to gather provisions, while simultaneously denying the same to the British. Moreover, Cornwallis was left "in doubt as to his own line of conduct" with American regular forces in close proximity to his lines of communication and major bases.[2]

Faced with the threat both Greene and Morgan posed, Cornwallis decided to pursue Morgan's light infantry corps—the greater threat to the forward base at Ninety Six. However, Cornwallis failed to coordinate the movement of his lumbering heavy infantry with Lieutenant Colonel Banastre Tar-leton's faster-moving *British Legion*. Furthermore, both British commanders underestimated Morgan's skill as a battlefield commander. As a result, Morgan cunningly lured Tarleton into attacking a care-fully prepared delaying defense at the Cowpens on 15 January 1781, which cost nearly 1,000 British dragoons and light infantry troops.

Deeply angered at the loss of most of his light infantry, Cornwallis abandoned all pretense of pacification in a futile attempt to destroy Greene's army and retake his prisoners (the Race for the Dan).

The British pursuit ended in failure on 14 February 1781 with Cornwallis's army marooned in a barren, unfriendly part of North Carolina, more than 200 miles (322 km) from the nearest forward base at Camden. The ragged American army's feat of arms was largely due to Greene's remarkable foresight and initiative. Greene painstakingly studied his area of operations and consequently was able to adapt his operations to make best use of the terrain. Furthermore, Greene gave himself un-matched operational mobility by minimizing baggage, and assembling a portable bridging company. As a result, Greene was able to ferry his entire army and logistics trains across the impassable Dan River to temporary safety in Virginia.

Vignettes

A good starting point for a campaign overview is to examine the leadership styles of the respective commanders. Greene cogently outlined his new strategy in a letter written shortly after taking com-mand of the Southern Department:

> I am here in my camp of repose, improving the discipline and spirits of my men. . . . I am
> well satisfied with the movement, for it has answered thus far all the purposes for which I

---

[2] William Johnson, *Sketches of the Life and Correspondence of Nathanael Greene, Major General of the Armies of the United States, in the War of the Revolution*, vol. 1 (Charleston, SC: A. E. Miller, 1822), 350–51.

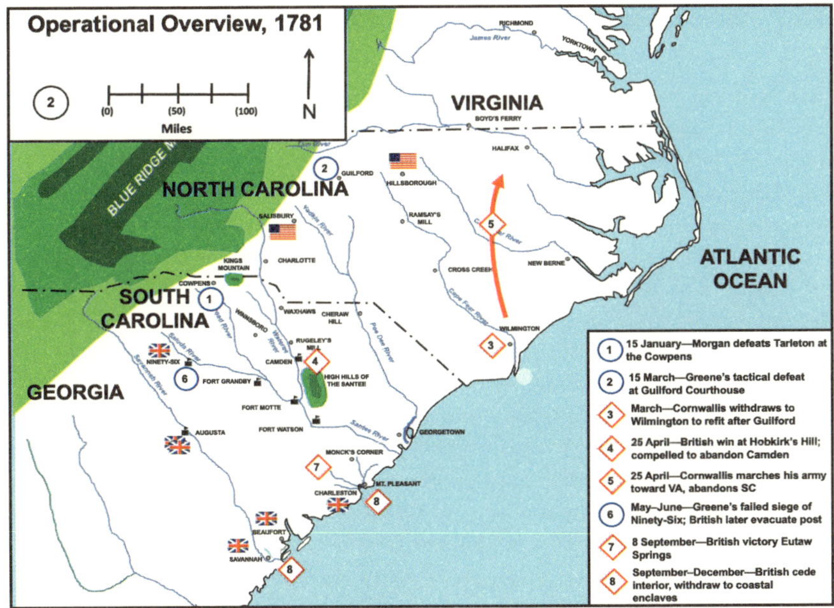

Map 2
Chuck Collins, Army University Press, and Harold Allen Skinner Jr.

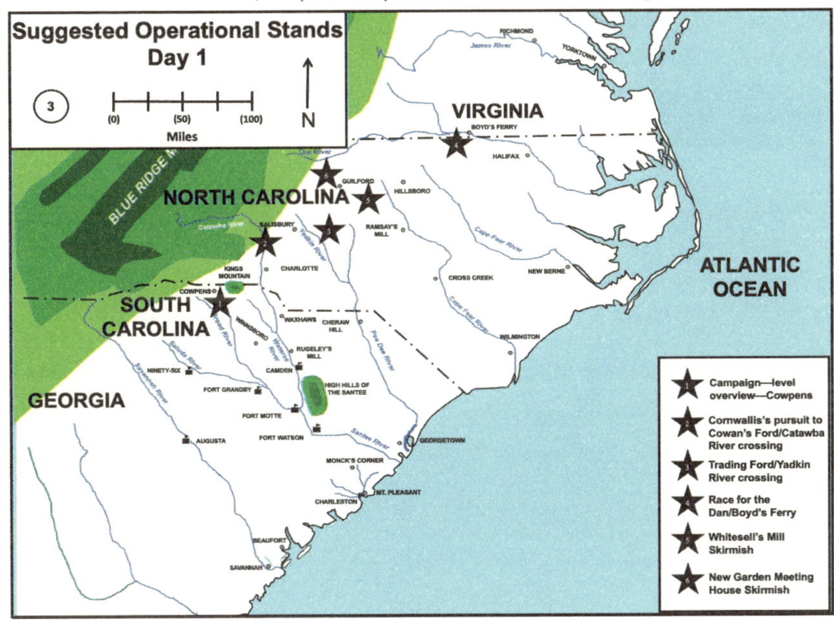

Map 3
Chuck Collins, Army University Press, and Harold Allen Skinner Jr.

Map Key

53

intended it. It makes the most of my inferior force, for it compels my adversary to divide his, and holds him in doubt as to his own line of conduct. He cannot leave Morgan behind him to come at me, or his posts of Ninety-Six and Augusta would be exposed. And he cannot chase Morgan far, or prosecute his views upon Virginia while I am here with the whole country open before me. I am as near to Charleston as he is, and as near to Hillsborough as I was at Charlotte, so that I am in no danger of being cut off from my reinforcements, while an uncertainty as to my future designs has made it necessary to leave a large detachment of the enemy's late reinforcements in Charleston. . . . It would be putting it in the power of my enemy to compel me to fight him. . . . I cannot venture to get entangled among the difficulties they present until I can turn upon my enemy and fight him when I please.[3]

Compared to his predecessors, Generals Benjamin Lincoln and Horatio Gates, Greene was a thoughtful, gifted, and self-confident commander, who seldom called a council of war to debate a course of action. Greene formulated his plans in private, using only trusted subordinates as a sounding board, and issued well-prepared mission-type orders. The change in command style made a deep impression upon the Southern Army, as Lieutenant Colonel Henry Lee noted:

> No man was more familiarized to dispassionate research than was General Greene. He was patient in hearing everything offered, never interrupted or slighting what was said; and having possessed himself of the subject fully, he would enter into a critical comparison of the opposite arguments, convincing his hearers, as he progressed, with the propriety of the decision he was about to pronounce.[4]

A postwar account from British commissary general Charles Stedman outlined how Greene's shift in strategy presented Cornwallis with a thorny dilemma.

> General Greene, who had succeeded Gates in the command of the American army, finding it difficult to procure a sufficient supply of provisions in the neighborhood of Charlotte . . . and being sensible that his present force was too weak to attempt any direct operation against lord Cornwallis; resolved to divide it, and by desultory incursions in different, and nearly opposite quarters, to alarm and harass the British out-posts on the frontiers of South Carolina. By such means his troops would be kept in action, and traversing different parts of the country, would not only be more easily supplied with provisions, but might in their process infuse some spirit into the militia, without whose assistance and co-operation he saw that he could do nothing effectual.[5]

Stedman then briefly outlined Cornwallis's operational plans after Cowpens:

> The disappointment was galling; and the loss of credit cast a shade over the commencement of the expedition. . . . The loss of the light troops [at the Cowpens] at all times necessary to any army . . . was not to be repaired. Deeply as his lordship was affected by the weight of this misfortune . . . he nevertheless resolved to prosecute the original plan of the expedition into North Carolina, as the only means of maintaining the British interest in the Southern colonies. . . . Some hopes were entertained that Morgan, incumbered as he was with prisoners, might still be overtaken between Broad River and the Catawba.[6]

---

[3] Johnson, *Sketches of the Life and Correspondence of Nathanael Greene, Major General of the Armies of the United States, in the War of the Revolution*, vol. 1, 350–51.
[4] As quoted in Theodore Thayer, *Nathanael Greene: Strategist of the American Revolution* (New York: Twayne Publishers, 1960), 296.
[5] Charles Stedman, *The History of the Origin, Progress, and Termination of the American War*, vol. 2 (London: J. Murray, 1794), 317–18.
[6] Stedman, *The History of the Origin, Progress, and Termination of the American War*, vol. 2, 325.

Analysis

1. Sir Henry Clinton had specifically ordered Cornwallis to not undertake any major offensive action until South Carolina and Georgia were pacified. How did Cornwallis's pursuit of Morgan and Greene meet or violate the letter and intent of Clinton's orders?

2. Consider General Greene's decisions immediately after the Cowpens and during the withdrawal across North Carolina. What were Greene's strengths and weaknesses? As you go through the subsequent stands, make sure to reevaluate your analysis: Does Greene shape the campaign, or was Greene shaped by events? A combination of both? What of Cornwallis?

Stand 2: Cornwallis's Pursuit, from the Cowpens to Cowan's Ford

Directions

From the visitor's center parking lot at the Cowpens National Battlefield, drive north to the intersection of South Carolina (SC) 11, and turn right (east). After 10 miles (16 km) you will arrive at the interchange with Interstate 85 (I-85). Take I-85 north approximately 50 miles (80 km) to Charlotte, North Carolina, before taking exit 30 on I-485 inner loop to the north and east. Travel on I-485 to exit 16 for North Carolina (NC) 16, proceeding north and west for around 14 miles (22.5 km). Turn right (east) on NC 73 and drive approximately 3 miles (4.8 km). Just before crossing the Catawba River turn left (north) into a parking lot marked as the McGuire Nuclear Power Plant Viewing Point. Park the vehicle and walk to the northeast corner (to the right) of the parking lot. If possible to do so safely, walk across the railroad tracks immediately to the front, and move as close as possible toward the river without trespassing on the power plant's property. For an excellent river level view of this stand, use the nearby Mountain Island Lake Highway 73 boat ramp. Simply drive past the McGuire viewing point parking lot, take an immediate left, and drive to the bottom of the parking lot. Walk east to the edge of the river, and conduct the stand as written.

An optional stand location is at a marker commemorating the death of General William Lee Davidson, the Patriot militia commander at Cowan's Ford. From the McGuire parking lot, turn left (east) on NC 73 and cross the Catawba River. About 500 meters east of the river is the intersection of Duke Power Cowan's Ford Road with NC 73. Immediately turn right (south) into a small pull-off area, and there will be a visible rough stone obelisk with a bronze tablet that commemorates General Davidson. Park and walk to the marker.

Orientation

The McGuire Viewing Point overlooks the general vicinity of Cowan's Ford, which in 1781 connected a secondary road that ran from Salisbury, North Carolina (about 40 miles [64 km] to the northeast), into western North Carolina. Modern engineering projects have made major changes to the topography that existed in 1780, and the actual location of Cowan's Ford is now submerged in the large reservoir clearly visible to the north of this location. After orienting students to the north, point to the 1300 position at the visible dam and spillway about 750 meters away, which marks the actual location of Cowan's Ford. About 4 miles (6.4 km) due north of Cowan's Ford was Beattie's Ford, the crossing point for the main east-west colonial road across the Catawba River. The battle site for the Cowpens is about 40 miles (64 km) to the southwest, while Charlotte, North Carolina, is about 20 miles (32 km) to the southeast. Finally, Ramsour's Mill, the location where Cornwallis's army burnt their logistics trains, is about 18 miles (29 km) to the west. Across the east bank of the river is the

Davidson marker, which was built in 1923 at the original site where General Davidson was mortally wounded. During the excavation for Lake Norman in 1971, the marker was removed from its original location overlooking Cowan's Ford and eventually relocated to its present location.

Description

In 1780, the Catawba River was a major obstacle to east-west travel across South Carolina, crossable only at places like Cowan's Ford. Immediately after the Battle of the Cowpens, Morgan's corps of about 1,000 troops, escorting 600 British and Loyalist prisoners, force-marched northeast from the battlefield, attempting to place the swift Catawba River between them and British pursuit. Meanwhile, Cornwallis's pursuit was delayed as Cowpens stragglers were policed up and the entire army reorganized with the arrival of 1,200 fresh reinforcements.

Lacking hard intelligence of Morgan's whereabouts, Cornwallis blindly marched toward the Cowpens until he learned of Morgan's location at Ramsour's Mill—60 miles (96.5 km) northeast of the Cowpens. By the time Cornwallis's army reached Morgan's old campsite on 24 January, the Americans were 50 miles (80 km) away, across the Catawba River. Cornwallis paused at Ramsour's Mill to rest and forage before destroying all nonessential baggage, supplies, and wagons, in an attempt to turn his heavy force into light infantry.[7]

While a rear guard marched the British prisoners toward Virginia, Morgan called out reinforcements from Brigadier General William Davidson's North Carolina militia to help cover the major Catawba fords. Davidson positioned 500 militia troops at Beattie's Ford and used companies of riflemen and dragoons to guard Cowan's Ford. Morgan positioned his corps at Sherrald's Ford, the direct route for Cornwallis's corps to approach the American logistics base at Salisbury, only 40 miles (64 km) to the northeast. Lacking tools and time, the American soldiers built log-and-branch obstacles and shallow fieldworks to cover the ford exits.

On 28 January, Cornwallis found the rain-swollen Catawba unfordable, with the major fords closely guarded by American skirmishers. That same day, General Greene and Brigadier General Morgan reconnoitered the defenses and concluded that reinforcements from Cheraw Hill, an 80-mile (128.7 km) march from Cowan's Ford, would not arrive in time to prevent Cornwallis's larger force from seizing a bridgehead on the east bank. As a consequence, Greene ordered the evacuation of Salisbury and its supplies into Virginia, screened by Morgan's delaying defense of the Catawba line. Despite the obvious risks of defeat, Greene remained optimistic, as news of Cornwallis burning his wagons meant the British corps was uniquely vulnerable to logistics culmination.[8]

By 1 February 1781, Cornwallis judged the Catawba shallow enough to force a crossing. Lieutenant Colonel James Webster's reinforced *33d Foot* created a diversion at Beattie's Ford, while the British main body attempted a predawn crossing at Cowan's Ford, led by a detachment of the *Brigade of Guards Light Infantry*, bearing unloaded muskets. Fire from the American pickets scattered Webster's Loyalist guides, forcing the redcoats to flail for the eastern bank. A large group, led by Lieutenant Colonel Francis Hall of the *Brigade of Guards*, blundered into deep water, and many troops were killed by drowning or enemy musket balls. Once the *Guards* reached dry land, a volley killed the luckless Brigadier General Davidson and forced the militia to flee from their positions. Cornwallis's report of the skirmish admitted to 4 men killed and 36 wounded, a total refuted by Patriot eyewitness accounts of many more dead British soldiers washed downstream of Cowan's Ford. By contrast, the Americans suffered few deaths aside from Davidson, although Tarleton claimed about 40 fighters killed and

---

[7] Angus Konstam, *Guilford Courthouse, 1781: Lord Cornwallis's Ruinous Victory* (Long Island, NY: Osprey Publishing, 2002), 42.

[8] Gen Greene to BGen Isaac Huger, 30 January 1781, in Dennis M. Conrad et al., *The Papers of General Nathanael Greene*, vol. 7, *26 December 1780–29 March 1781* (Chapel Hill: University of North Carolina Press, 1994), 220.

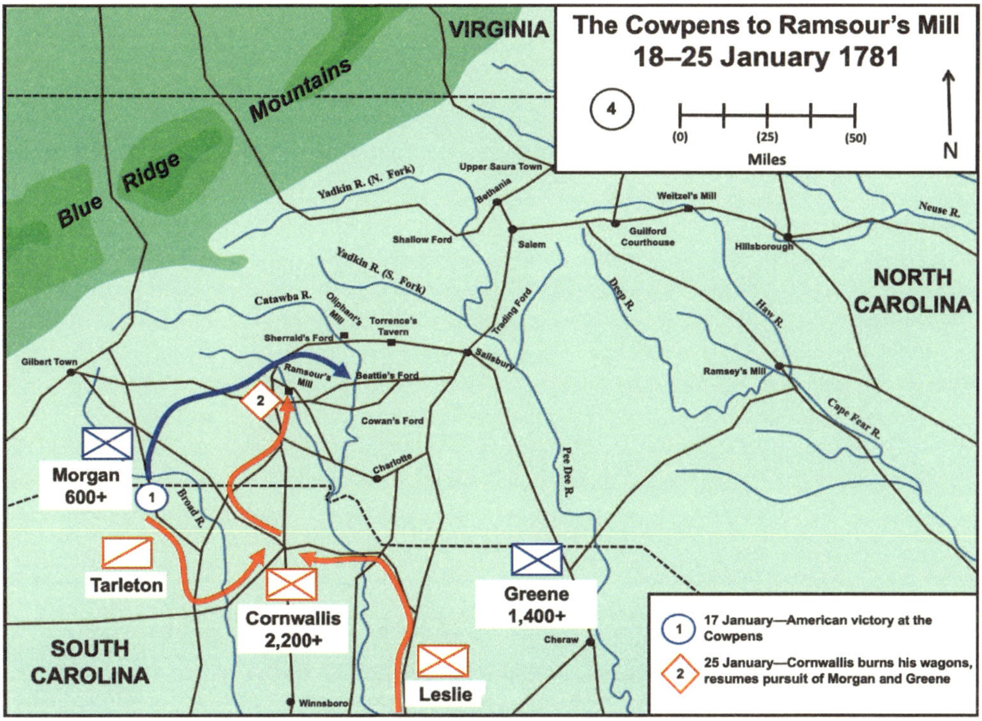

Map 4
Chuck Collins, Army University Press, and Harold Allen Skinner Jr.

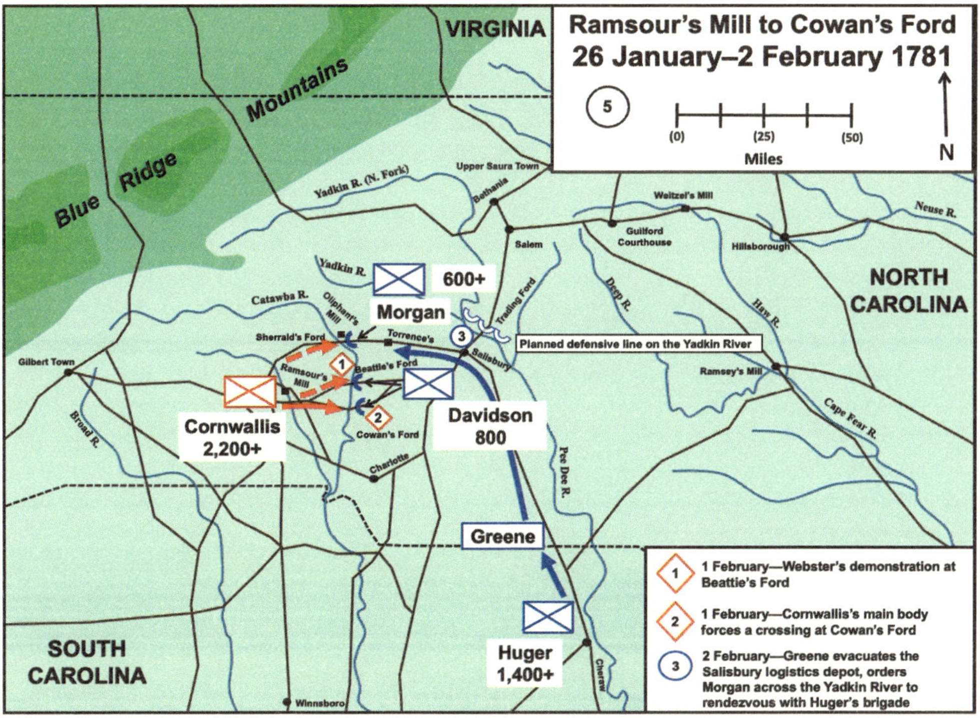

Map 5
Chuck Collins, Army University Press, and Harold Allen Skinner Jr.

wounded. Casualties aside, Cornwallis's superb light infantry had successfully breached a defended water obstacle and unhinged the American defenses on the Catawba, forcing the Americans into a race to evacuate Salisbury before the British could exploit the situation.[9]

Vignettes

Letters written by Brigadier General Daniel Morgan illustrate how he planned and deployed his area defense, despite suffering near-debilitating pain from chronic sciatica and rheumatism:

> I arrived here this morning. The prisoners crossed at the Island Ford seventeen miles higher up the river. . . . Shall send them on to Salsbury in the morning. . . . Lord Cornwallis, whether from bad intelligence or to make a show, moved up toward Gilbert Town to intercept me the day after I had passed him. . . . I intend to stay at this place till I hear from you. In order to recruit the men and to get in a good train, we must be fited out with pack horses for as I wrote you before waggons will not do for light Troop. I intend to send Col Pickens back immediately in order to keep up a show of opposition, and to cut of[f] small parties that may be sent out . . . to prevent us from getting supplies In that back country.[10]

> Lord Cornwallis on the 24th Instant Encamped at Ramsowers Mill, with his Main Body on his way from Broad River. His Advanced corps Moved Eight Miles further this way In the Night, and returned the Next day to their Main Body. He Still continues at that place. . . . . I am Trying to Collect the Militia, to Make a Stand at this place. Genl [William L.] Davidson with five hundred militia, two hundred and fifty of which are without flints, I have ordered to Beaties ford. We are filling all the Private fords so as to Make them impassable. The one that I Lie at I intend to Leave Open. On Lord Cornwallis' approach I thought it advisable to order all the Prisoners and Stores from Salisbury towards the Moravian Town. . . . I am a Little Apprehensive, that Lord Cornwallis Intends to Surprise me, lying so Still this day or two, but if the Militia dont Decieve me . . . I Think I will put it out of his Power. . . .[11]

> Beaties Ford 29 Jan 1781. Sir, I just arrived at this place to view the situation. Gen. Davidson is here with Eight hundred men. The enemy is within ten miles of this place in force, their advance is in sight. It is uncertain whether they intend to cross here or not. I have detached two hundred men to the Tuckaseega Ford to fill it up & Defend it. An express just arrived who informs they have burned their wagons and loaded their men very heavy. We have taken four prisoners who says they are for Salisbury. I am Just returning to Sherald Ford where our regulars lie. I expect they will attempt to cross in the morning.[12]

Commissary General Stedman's report of the breaching of the Catawba defenses related:

> In the course of two days, the river having fallen so as to render it fordable, lord Cornwallis determined to attempt a passage. . . . lieutenant-colonel Webster, with one division of the army, was detached to a public ford called Beattie's . . . and make a feint . . . whilst lord Cornwallis, with the other division, marched to a private ford near M'Cowan's . . . The numerous fires seen on the opposite shore quickly convinced the British commander that this ford . . . had not escaped the vigilance of the enemy. . . . Nevertheless, lord Cornwallis determined

---

[9] Babits and Howard, *Long, Obstinate, and Bloody*, 22.
[10] Conrad et al., *The Papers of General Nathanael Greene*, vol. 7, 178.
[11] Conrad et al., *The Papers of General Nathanael Greene*, vol. 7, 211.
[12] Conrad et al., *The Papers of General Nathanael Greene*, vol. 7, 215.

to proceed . . . by the brigade of guards under general O'Hara. Plunging into the rapid stream, in many places reaching above their middle, and near five hundred yards wide, they marched on with the utmost steadiness and composure; and although exposed to the fire of the enemy, reserved their own . . . until they reached the opposite bank. . . . When the light-infantry had nearly reached the middle of the river they were challenged by one of the enemy's centinels . . . by discharging his musquet; and the enemy's picquets were turned out. . . . Colonel Hall, being forsaken by his guide . . . led the column directly across the river . . . This direction . . . carried the British troops considerably above the place where the ford terminated on the other side, and where the enemy's picquets were posted . . . When general Davidson perceived the direction of the British column, he led his men to that part of the bank which faced it. But . . . the light-infantry. . . . quickly routed and dispersed general Davidson's militia. . . . General Davidson . . . received a mortal wound. . . . the whole loss of the guards amounted only to forty, lieutenant-colonel Hall, and three privates, being killed, and thirty-six wounded.[13]

Analysis

1. What were the pros and cons to Cornwallis's decision to burn his logistics trains and equipment? What risks did he assume? What could Cornwallis have done to mitigate those risks?
2. What assumptions did Greene make when learning of Cornwallis's decision?
3. Analyze the performance of both sides during the Catawba phase of the campaign. What were the British strengths and weaknesses during this phase? What could have been improved in the American defense? Was Greene's decision to fight a delay versus a deliberate defense a good choice? Why or why not?

Stand 3: Cowan's Ford to Trading Ford (Yadkin River) 2–6 February 1781

Directions

From the McGuire Nuclear Power Plant Viewing Point, turn left (east) on NC 73. Follow NC 73 for approximately 18 miles (28.9 km) to the intersection with I-85 in Concord, North Carolina. Take I-85 north 27.8 miles (44.7 km) to exit 84 (Tributary Way) in Linwood, North Carolina. At the base of the off-ramp, turn left (north) on NC 150 for 0.4 mile (0.6 km) and turn left (west) on US 29 S/US 70 W. After about 0.5 mile (0.8 km), turn right (north) on County Road 1139 (Sowers Road) and take an immediate left (west) at the first crossroad, which on some maps is called Old Salisbury Road; on newer maps, the road is named Trading Ford Way. Follow Trading Ford Way about 0.7 mile (1.1 km) to the York Hill Yadkin River Access. There are two options for viewing this stand. For a rivers-edge view, drive down the hill following signs for the boat ramp. Use caution in driving up or down as the grade is a bit steep for low-clearance vehicles. Adjacent to the boat ramp is a sizable gravel parking lot. Park in a location away from boat trailers and walk west to the edge of the Yadkin River. For an excellent birds-eye view of the Yadkin and its adjoining terrain, albeit with the risk of road noise from the adjacent US 29 roadbed: after turning on US 29 S/US 70 W, drive 0.7 mile (1.1 km) and slow down to look for an unmarked turn-off to the right. Use caution; if the turn is missed, the driver will have to drive across the Yadkin River to safely make a U-turn. The turn-off ends in a parking lot with

---

[13] Stedman, *The History of the Origin, Progress, and Termination of the American War*, vol. 2, 327–29.

room for only two 15-passenger vans or four smaller vehicles. Park and walk west about 100 meters out on the bridge span to a point overlooking a small island in the Yadkin River.

Orientation

This location overlooks the approximate location of Trading Ford, a major trade route crossing point over the Yadkin River. From this point, Cowan's Ford is about 50 miles (80.4 km) to the southwest, while the major Continental logistics base at Salisbury was only 8 miles (12.8 km) to the southwest. Guilford Courthouse is approximately 41 miles (66 km) to the northeast, while Charlotte is about 42 miles (67.6 km) to the southwest.

Discussion

During the 1780s, the major road between Salisbury and the village of Salem (modern Winston -Salem) crossed the Yadkin River at Trading Ford, which was in the general location of this stand. This location serves as a good point to discuss the next major phase of Cornwallis's pursuit operation. After the Catawba crossing, Cornwallis briefly rested his drenched soldiers at a campsite 6 miles (9.6 km) east of Beattie's Ford on the night of 1–2 February 1781. Correctly assuming that Morgan had ordered his troops to withdraw from the Catawba, Cornwallis ordered Tarleton's *Legion* to pursue and regain contact with the Americans. Tarleton's interrogation of locals revealed that a dispirited (and possibly drunk) contingent of Davidson's North Carolina militia had assembled at the small crossroads of Torrence's Tavern, about 12 miles (19.3 km) from Cowan's Ford. Tarleton drove his fighters hard, and in short order the Patriot fighters and many of their families were routed in a quick skirmish that left several Americans dead in the road. Tarleton's troops followed up their victory by plundering and burning the tavern and abandoned wagons left at the crossroads. Taken together, the death of Brigadier General Davidson and the defeat at Torrence's Tavern was enough to prompt the local Patriot militia to entirely quit the campaign.[14]

After receiving Morgan's report of the collapse of the Catawba defenses, Greene immediately ordered the Continentals and wagon trains to withdraw to the next defensible terrain feature, the Yadkin River. Greene replied to Morgan, with orders to break contact and withdraw his light troops to Salisbury. After a hasty rendezvous with Morgan at Salisbury, Greene personally oversaw the evacuation of the Continental logistics depot, with its supplies, skilled technicians, and factories, toward the Trading Ford. Since Cornwallis was already east of the Catawba, Greene sent a fragmentary order to Brigadier General Isaac Huger's Virginia Continental Brigade and Lieutenant Colonel Henry Lee's Partisan Legion, directing a consolidation at the Trading Ford. Greene and Morgan arrived at Trading Ford early on 3 February to find the rain-swollen Yadkin near flood stage. Greene judged that the risk of crossing was preferable to defending with a water obstacle to the rear, assuming Cornwallis would push toward the ford without rest. Bateaux loads of Continental troops, supplies, and materials crossed the Yadkin throughout the day of 3 February. Once the Continentals were across the river, the militia rearguard swept the west bank of the Yadkin clear of boats before disbanding and returning home.

Cornwallis paused his corps at Salisbury only long enough for his soldiers to gather leftover supplies yet did not reach the deserted Trading Ford until midnight on 3–4 February. Despite having lost a second race for a river crossing, primarily due to bad intelligence and supply shortages, Cornwallis refused to call off the pursuit. Instead, the British troops camped at Trading Ford, and searched for increasingly scarce provisions while waiting for the Yadkin to drop to a fordable level. Safely across

---

[14] Babits and Howard, *Long, Obstinate, and Bloody*, 24.

the river, the scattered American forces converged on Guilford Courthouse, while General Greene remained near Trading Ford, to oversee the American rear guard. When Greene observed the British break camp and march north on 5 February, he correctly assumed Cornwallis was looking for a suitable crossing point to resume the pursuit. Also, Cornwallis's advance played into Greene's targeting of supplies and support. Every step northward brought Cornwallis's command closer to its logistics culmination point. As Greene had noted: "From Lord Cornwallises pushing disposition, and the contempt he has for our Army, we may precipitate him into some capital misfortune."[15]

Vignettes

Cornwallis's summary of the Catawba breaching operation, in a letter to Lieutenant Colonel Lord Francis Rawdon-Hastings at Charleston:

> We passed the Catawba on the 1st at a private ford, about four miles below Beatty's. The Guards behaved gallantly, and, although they were fired upon during the whole time of their passing by some militia under General Davidson, never returned a shot until they got out of the river and formed. . . .
>
> I am much distressed by the rivers and creeks being swelled, but shall try to pass the Yadkin at the shallow ford as soon as possible. I have the utmost confidence in your abilities and discretion. Our friends must be so disheartened by the misfortune of the 17th that you will get but little good from them. You know the importance of Ninety-Six: let that place be your constant care.[16]

Unlike Cornwallis, Greene had to devote considerable attention to matters concerning his militia, as was reflected in a letter written to General Andrew Pickens:

> As you have but few men with you here, and as you may have it in your power to collect considerable numbers in the rear of the enemy you will repair there immediately, and draw together as large a force as possible, and employ it in harrassing them on their march, and in preventing their parties from being sent out after forage. You will advise me of the movements of the enemy, and of the force you collect. It will be attended with many advantages to collect a force on the other side of the Catabawa. For this purpose you will detach three or four good officers to raise the Militia in that quarter; and to have them in readiness to be employed as ocasion may require. I shall write to General Sumter to pursue similar measures; and you will please to correspond with him on the subject, that the whole collective force may be employed to one point.
>
> You must take the usual methods, pointed out by the Laws of the State for supplying your Men with provisions, forage, horses and all other matters necessary for furthering the operations; but if these should fail and the emergencies of service render more decisive measures necessary, you must impress, taking care that it is done by Officers, and that proper certificates are given for the things taken. You will be watchful and vigilant to guard against a surprise and have your men as much collected as possible; and in order to reduce matters to a greater certainty and more order it would be well to engage the men for some given time, that they shall not leave Camp without leave. Let the troops you collect be as lightly equiped as possible, that you may move with expedition to join this army. . . . If you

---

[15] John Buchanan, *The Road to Guilford Courthouse: The American Revolution in the Carolinas* (New York: Wiley, 1997), 350.

[16] Charles Ross, ed., *Correspondence of Charles, First Marquis Cornwallis*, vol. 1 (London: John Murray, 1859), 83–84.

act in conjunction with the North Carolinia Militia, and there is no officer of equal rank, you will assume the command, until a more perfect arrangment can be made.[17]

Analysis

1. After performing a hasty analysis of mission, enemy, terrain and weather, troops and support available, time available, and civilian considerations, consider if General Greene should have reasonably foreseen the failure of the Catawba defense? What are some possible measures he could have taken to improve the odds?
2. Despite his outspoken criticism with the performance of the militia, Greene had no choice but to depend on militia support to reinforce his Continentals. Analyze Greene's use of the militia at the Catawba in comparison with later operations, specifically the Whitesell's Mill and the Guilford Courthouse first line fight. What, if anything, did he do better? How did the American militia's perceptions of their treatment by Greene affect the course of the campaign?
3. Cornwallis encountered many problems during this phase of the campaign due to faulty intelligence. How could Cornwallis have improved the quality and quantity of intelligence?
4. What are some of the ways General Greene attempted to influence the civilian population during this phase of the campaign? What was he trying to accomplish? How well did he succeed?

Stand 4: Crossing the Yadkin and the Race for the Dan River, 6–15 February 1781

Directions

From the Yadkin River overlook, drive south on Old Salisbury Road to US 70W/US29 S/NC-150W. Take US 70E/US 29N 4.6 miles (7.4 km) to the interchange with I-85, and take the northbound ramp toward High Point/Greensboro, North Carolina. Continue on I-85 to exit 223 for US 29N/US 70E/US 220N toward Reidsville, North Carolina. Follow US 29N until it merges with US 59E toward Lynchburg/South Boston, Virginia. Exit on US 360E/US 58E toward South Boston, and travel on US 360E/US 58E to Main Street. In South Boston, turn left (north) on Main Street, which turns into Broad Street at the end of the 1-km drive across the bridge spanning the Dan River. Make an immediate left on Factory Street after crossing the bridge. Follow the curve of the road northeast about 180 meters. Turn left (west) on Seymour Street, and remain on Seymour as the road curves to the southeast. Cross the railroad tracks, and follow the brown Boyd's Ferry sign to the left (southeast) about 170 meters. To the northwest of two small brick buildings is a small parking spot sufficient for several 15-passenger vans. The closest parking suitable for a bus-size vehicle is 0.5 miles (0.8 km) to the northwest, at the trailhead of the Tobacco Heritage Trail. If the Heritage Trail location is used for parking, it is recommended to drop students at the Boyd's Ferry location as the road between is not suitable for large groups of pedestrians. Note: the remaining 100 meters or so of unimproved road from the brick buildings to the banks of the Dan River are in a flood-prone river bottom, so a foot reconnaissance is recommended before attempting a vehicle movement beyond this point. However movement is done, follow the unimproved road from the brick buildings to the southeast until the bank of the Dan River becomes visible.

---

[17] Conrad et al., *The Papers of General Nathanael Greene*, vol. 7, 241–42.

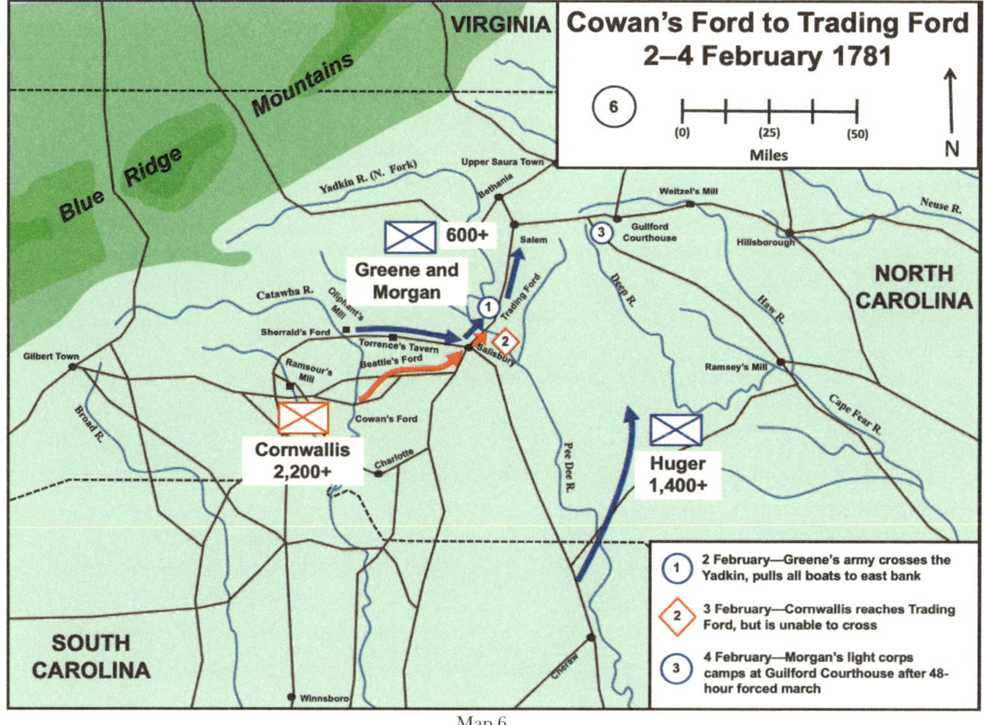

Map 6

Chuck Collins, Army University Press, and Harold Allen Skinner Jr.

Orientation

This location on the Dan River overlooks the general location of Boyd's Ferry, the principal crossing point used by the Continental Army to reach its Virginia sanctuary. While looking directly (southeast) at the riverbank, point to the right (southwest), the general direction of Guilford Courthouse, about 77 miles (124.9 km), and Trading Ford on the Yadkin, about 120 miles (193 km). Upstream to the right (west) about 5 miles (8 km) was Irwin's Ferry, the other major crossing point used by Greene's army.

Description

During the Revolution, Boyd's Ferry was the principal logistics trans-shipment point for the Southern Department, thus a logical crossing point for Greene's army into Virginia. About 25 meters from the road stands a white granite obelisk and replica bateau (useful as a teaching prop) to commemorate the crossing. Bateaux were flat-bottomed boats up to 35 feet long, propelled by four to six oarsmen, suitable for carrying payloads up to 0.75 tons along shallow rivers.[18]

When Greene learned of Cornwallis's crossing of the Yadkin River on 5 February, the Continental forces were ordered to regroup at Guilford Courthouse. During an 8 February muster at the courthouse, Lieutenant Colonel (and Adjutant General) Otho Williams found just 1,400 Continentals and 800 militia present for duty. A council of war reaffirmed Greene's notion to retrograde across the Dan River. Here, the ailing Morgan was replaced by Lieutenant Colonel Williams, second in command of the veteran 1st Maryland Continental Regiment. To better handle rearguard tasks, Williams's flying corps was regrouped to incorporate all light infantry and dragoon units.

---

[18] Carol V. Ruppé and Janet F. Barstad, eds., *International Handbook of Underwater Archaeology* (New York: Springer Science and Business Media, 2002), 79.

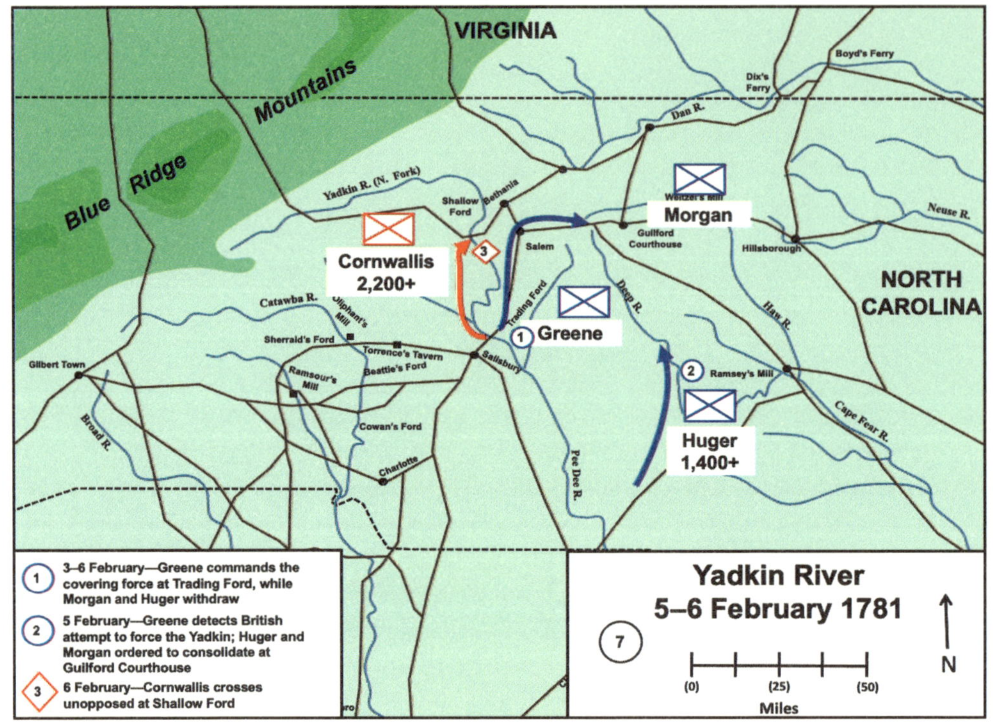

Map 7
Chuck Collins, Army University Press, and Harold Allen Skinner Jr.

Once across the Yadkin, Cornwallis marched toward the upper Dan River, based on intelligence that the lower Dan fords were impassable at high water. En route, Cornwallis's army marched through several Moravian villages, where the hungry British and Hessian soldiers pillaged corn and more than 90 head of cattle, plus smaller livestock.[19] Cornwallis also met with the exiled Royal governor Josiah Martin, but as Martin lacked widespread legitimacy among the civilian population, he was unable to offer significant support to Cornwallis's efforts.

Unlike Cornwallis, Greene did not have to worry about river depths due to his efficient bateaux train commanded by Quartermaster General Lieutenant Colonel Edward Carrington. To streamline the crossing, half the army marched to Boyd's Ferry and the remainder toward Irwin's Ferry. Williams was ordered to stage a diversion by marching toward the upper Dan River, thus screening the withdrawal of the slower main body. Greene's orders to Williams were specific: delay by force without becoming decisively engaged and cross the Dan River only after the main body had reached sanctuary in Virginia.[20]

Today, the width of the Dan River at Boyd's Ferry, near South Boston, Virginia, averages 82 yards (75 meters) in width, 10 feet in depth, with a 4-knot-per-hour flow rate. At a flood stage of 19 feet, the river flows at about 14 knots per hour, probably similar to the flow rate of February 1781. Given those conditions, Greene was certain that Cornwallis's army would find the Dan River a temporarily impassable obstacle, especially after Carrington's men swept all boats from the area.[21]

---

[19] Babits and Howard, *Long, Obstinate, and Bloody*, 27. The Moravians were a pacifist sect that had settled in the region on a 1753 grant of land from the king that included an exemption from military service. Often derided by both sides for their pacifism, the Moravians were an invaluable source of supplies, funding, and medical care for Greene's army. See Ruth Blackwelder, "The Attitude of the North Carolina Moravians Towards the American Revolution," *North Carolina Historical Review* 9, no.1 (January 1932): 5, 16–17, 20, for more details.

[20] Babits and Howard, *Long, Obstinate, and Bloody*, 32–35.

[21] For hydrographic results, see "Advanced Hydrologic Prediction Service: Dan River at South Boston," National Weather Service (website), accessed 28 May 2018.

After blundering about trying to split Huger and Greene's dispersed elements, Cornwallis learned of Greene's actual location at Guilford Courthouse, roughly 25 miles (40 km) to the northeast. News of Cornwallis's turn toward Guilford Courthouse prompted Greene to resume his army's withdrawal into Virginia.[22] Williams's deception movement toward the upper Dan River drew Cornwallis's corps northward in expectation of a decisive battle of annihilation.

The tempo of Cornwallis's pursuit was unrelenting, with movement starting before daybreak and constant skirmishing until the end of dusk. In one instance, Tarleton's troopers surprised Lee's troops taking a hasty breakfast halt. Abandoning their hot breakfast, Lee's soldiers fought a brief running skirmish before escaping by ripping the planks from a bridge. The lack of hot food and serviceable uniforms and blankets added to the cumulative effects of cold weather, combat stress, and fatigue on the troops of both armies.

Williams's delaying defense allowed Greene's main body and wagon trains to safely cross the Dan River during the night of 13–14 February. Once the crossing was complete, Greene ordered Williams to break contact and drive toward safety. The receipt of Greene's orders bought a loud cheer from Williams's men, unmistakable evidence that Greene had once again outgeneraled Cornwallis. Screened by Lee's dragoons, Williams's invigorated infantry broke contact and safely crossed at Boyd's and Irwin's Ferries late on 14 February. At sunrise the next day, Tarleton's dragoons found at Irwin's Ferry "some works evacuated, which had been constructed to cover the retreat of the enemy, who six hours before had finished their passage, and were then encamped on the opposite bank. . . . Every measure of the Americans, during their march from the Catawba to Virginia, was judiciously designed and vigorously executed."[23]

Vignettes

On 5 February, Greene issued a fragmentary order redirecting Brigadier General Isaac Huger's brigade to Guilford Courthouse.

> Dear Sir
>
> I wrote you yesterday of our situation, and the enemies at Tradeing Ford. . . . We found the river was falling last night so fast, that it might be forded this morning; and as our force was too small and the ground unfavorable for defending the fords, it was thought most advisable to retire last evening. We are on our march for Guilford. It is our interest to form a junction as soon as possible; but it is not absolutely certain, that the enemy will cross the Country for Guilford. . . .
>
> I intend to try to collect the Militi[a] about Guilford if possible; and if we can find a good position, prepare to receive the enemies attack. It is not improbable from Lord Cornwallises pushing disposition, and the contempt he has for our Army, we may precipitate him into some capital misfortune. I wish to be prepard, either for attacking, or for recieving one. . . .
>
> If Lord Cornwallis knows his true interest he will pursue our army. If he can disperse that, he compleats the reduction of the State: and without that he will do nothing to effect.[24]

Continental Sergeant Major William Seymour's account of the retrograde:

> On the eighth instant we marched from here [Guilford Courthouse] . . . taking one road and the light troops another. . . . This day we received intelligence that the British Army

[22] Babits and Howard, *Long, Obstinate, and Bloody*, 32.
[23] Babits and Howard, *Long, Obstinate, and Bloody*, 35.
[24] Conrad et al., *The Papers of General Nathanael Greene*, vol. 7, 251. Salt was a vital commodity in the days before canning and refrigeration, as the mineral was used to preserve meat and fish.

was advancing very close in our rear, upon which Colonel Lee detached a party of horse to intercept them, who meeting with their vanguard, consisting of an officer and twenty men, which they killed, wounded and made prisoners. . . . We . . . reached [the Dan] on the fourteenth, after a march of two hundred and fifty miles from the time we left our encampment at Pacolet River [before the battle of the Cowpens]. By this time it must be expected that the army . . . were very much fatigued both with travelling and want of sleep . . . we marched for the most part both day and night, the main army of the British being close in our rear, so that we had not scarce time to cook our victuals.[25]

British Commissary Stedman described the movements after crossing the Yadkin:

All the boats and flats having been secured by Morgan on the other side of the river, the river itself being unfordable . . . lord Cornwallis determined to march up the western banks of the Yadkin and pass by the shallow fords. . . . All hopes of preventing the junction of the two divisions of the American army were now at an end; but still another object . . . remained . . . to get between the American army and Virginia, to which province it was obvious general Greene meant to retreat. . . . The lower fords [of the Dan] were represented to be impassable in the winter season, and . . . a sufficient number of flats could not be collected to transport the American army. . . . Lord Cornwallis, misled by this information, directed his march to the upper fords upon the Dan, in order to intercept Greene's retreat. . . . and so much had lord Cornwallis been misinformed . . . that the American troops . . . with their baggage . . . were passed over with ease, at Boyd's and Irwin's Ferries, in the course of a single day. . . . The light army . . . was so closely pursued, that scarcely had its rear landed, when the British advance appeared on the opposite banks. . . . The hardships suffered by the British troops . . . in this long and rapid pursuit . . . were uncommonly great; yet such was their ardour in the service, that they submitted to them without a murmur. . . . And that the latter escaped without suffering any material injury, seems more owing to a train of fortunate incidents, judiciously improved by their commander, than to any want of enterprise or activity in the army that pursued.[26]

Cornwallis wrote an operational summary of the pursuit.

I tried by a most rapid march to strike a blow either at Greene or Morgan before they got over the Dan, but could not effect it. The enemy, however, was too much hurried to be able to raise any militia in this province. The fatigue of our troops and the hardships which they suffered were excessive. I receive strong assurances from our friends. To-morrow the King's standard will be erected, and I shall try every means to . . . avail myself of their services. I cannot be sure when I shall be able to open the communication with Cross Creek; it must be done soon as the troops are in the greatest want of shoes and other necessaries. As I am informed that Greene expects reinforcements . . . and that Virginia militia are turning out . . . in great numbers, I should wish the three regiments expected from Ireland to be sent to me as soon as possible by way of Cape Fear . . . to Cross Creek.[27]

[25] William Seymour, "A Journal of the Southern Expedition, 1780–1783," in *The Pennsylvania Magazine of History*, vol. 7 (Philadelphia: Historical Society of Pennsylvania, 1883), 296.
[26] Stedman, *The History of the Origin, Progress, and Termination of the American War*, vol. 2, 331–32.
[27] Charles Ross, *Correspondence of Charles, First Marquis, Cornwallis*, vol. 1 (London: John Murray, 1859), 84. Cross Creek is near modern day Fayetteville, NC. In January 1781, a small British corps had seized Wilmington as a staging base to send supplies and reinforcements via the Cape Fear River to Cornwallis at Cross Creek. Lindley S. Butler and John Hairr, "Wilmington Campaign of 1781," NCPedia (website), 2006.

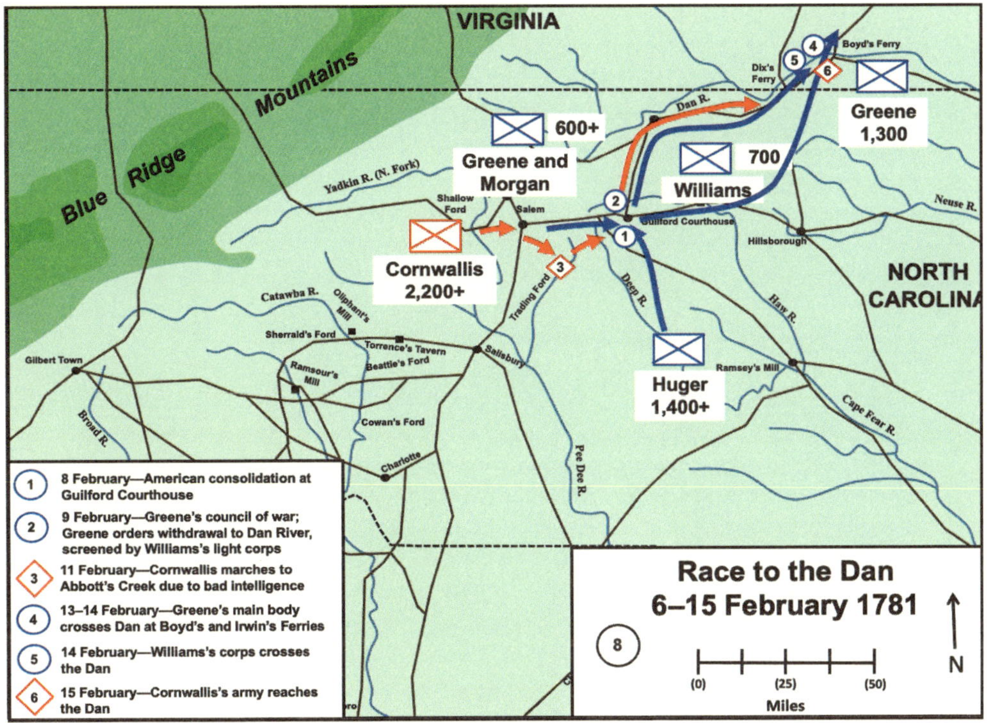

Map 8
Chuck Collins, Army University Press, and Harold Allen Skinner Jr.

Analysis

1. What were General Greene's principal reasons to withdraw into Virginia? Which risks did he avoid, and which did he assume? How did he mitigate those risks?

2. Why was a counterattack not a viable option for Greene's army at this point?

3. On the surface, Cornwallis successfully ends the Race for the Dan by running Greene's army from the Carolinas. Did Cornwallis accomplish anything of long-term significance? Furthermore, what were his options at this point in time?

4. Cornwallis's letter of 21 February indicates his concerns with reinforcements and resupply. *Sustainment Operations*, Field Manual (FM) 4-0, paragraph 5-94, notes in part "a successful sustainment plan will extend operational reach, prevent culmination or loss of the initiative, manage transitions, exploit possible opportunities, and mitigate risk." Marines may refer to chapter 13 "Sustainment Operations" in *Marine Corps Operations*, Marine Corps Doctrinal Publication (MCDP) 1-0 (w/Change 1).[28] At this point in the campaign, what could Cornwallis have done to reduce the risk of logistics culmination?

Stand 5: The Battle of Whitesell's Mill, 6 March 1781
Directions

From the location overlooking Boyd's Ferry, return to your vehicle. Drive south across the Dan River

---

[28] *Sustainment Operations*, FM 4-0 (Washington, DC: Department of the Army, 2019), 5-23; and *Marine Corps Operations*, MCDP 1-0 (w/Change 1) (Washington, DC: Headquarters Marine Corps, 2017), 13-1–13-3.

bridge on Main Street and turn right (west) on US 360E/US 58E/US 29S heading southwest toward Greensboro, North Carolina. Follow signs to remain on US 29S to the south and west of Danville, Virginia. Inside North Carolina, remain on US 29S and drive until across the Haw River, after which is an intersection with US 29 BUS. Turn south on US 29 BUS, then east (left) onto Candy Creek Road, followed by an immediate right turn to the south onto County Road (CR) 2629 (Friendship Church Road). Follow CR 2629 for about 1 mile (1.6 km) to a side intersection with CR 2703 (Chrismon Road). Take Chrismon Road southeast (left) to the intersection with NC-61 and NC-150. Follow signs to stay on NC-61S for approximately 8 miles (12.8 km). At the side intersection with CR 2734 (Bellflower Road) begin watching the odometer. At about 0.4 mile (0.6 km), you will cross a bridge over Reedy Fork Creek, which for electronic mapping purposes is located at 36.179259N–79.576374W. Slow the vehicle and carefully pull to the left (southeast) onto the grassy shoulder, which has sufficient room for up to three 15-passenger vans. Park and move by foot toward the right side of the bridge. Caution in descending to the creek is in order due to the steep and rocky slopes. Furthermore, the slope is covered with poison ivy so participants should take appropriate precautions.

A good alternative to the bridge site is located at the end of Woellner's Way Road, which branches off NC-61 about 350 meters north of the bridge. From NC-61, follow Woellner's Way about 500 meters as it curves to the south and east before ending at a concrete abutment overlooking Reedy Fork Creek. This location has sufficient room to parallel park two 15-passenger vans. Clearly visible about 100 meters to the southeast is the foundation of Whitesell's Mill. Walk about 60 meters to the west bank of Reedy Fork Creek. **CAUTION:** the end of the road and adjacent land are part of the private property at 6401 Woellner's Way, Gibsonville, North Carolina. Prior to using this stand, the facilitator **must obtain permission** from the property owner for use of the road and land adjacent to the mill.

Orientation

This location overlooks where Lieutenant Colonel Lee fought a delaying engagement against Lieutenant Colonel Webster's brigade on 6 March 1781, the largest skirmish to take place after Greene's recrossing into North Carolina, and the Battle of Guilford Courthouse on 15 March. After orienting students to the north, point out NC-61 as the approximate trace of the Revolutionary-era road between Whitesell's Mill and High Rock Ford to the north. Roughly beneath the roadbed of the modern bridge was the Horse Ford, while the main ford was in the sharp curve about 250 meters to the south of this location. Next, point out the stone foundations of Whitesell's Mill, remnants left when the mill was destroyed by severe flooding. Overlooking the main ford was a sturdy log building that in prewar times served as an Alamance County schoolhouse and served as an American strongpoint during the battle. Heinrich Whitesell's farmstead was about 250 meters to the northwest of the mill. To the southeast about 13 miles (21 km) was Cornwallis's camp along Alamance Creek, while Greene's main camp was near High Rock Ford on the Haw River, about 8 miles (12.8 km) due north.

Discussion

Greene's withdrawal into Virginia mitigated his supply situation, allowing time for pressing the governors of North Carolina and Virginia for fresh Continental recruits. On 19 February 1781, Lee's Partisan Legion was reorganized as a corps of observation with attached Continental and militia troops and sent back into North Carolina. There Lee was to aggressively gather intelligence about Cornwallis's strength and dispositions and prevent British consolidation efforts.[29] Meanwhile, Corn-

---

29 Buchanan, *The Road to Guilford Courthouse*, 359–60.

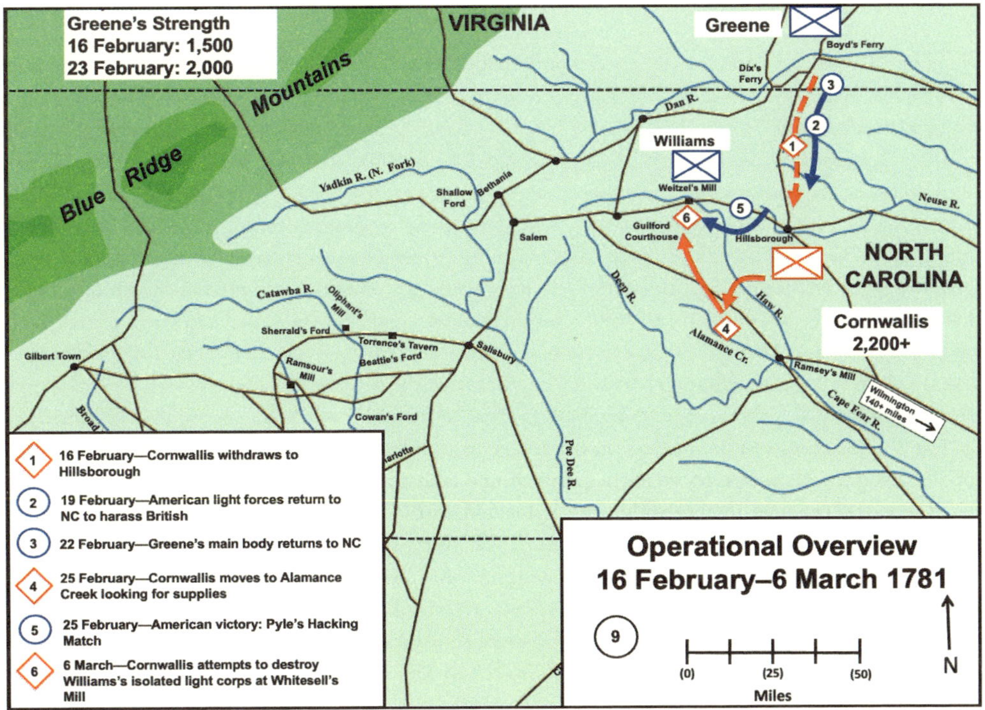

Map 9
Chuck Collins, Army University Press, and Harold Allen Skinner Jr.

wallis had marched to Hillsborough, North Carolina, where Commissary Stedman had gathered supplies by deploying infantry patrols to take livestock and provisions from known Patriot sympathizers.[30]

Buoyed by the presence of the army, many Loyalists arrived to help resupply efforts and enroll in new Loyalist militia units. Recruiting efforts quickly ground to a halt when word spread of the destruction of Colonel John Pyle's Loyalist militia battalion at the hands of Lieutenant Colonel Lee's Legion.

While Cornwallis struggled to regroup at Hillsborough, Greene's army was well-provisioned and gained strength with two brigades of Virginia State Line troops and two Virginia rifle battalions. Suitably reinforced, Greene crossed into North Carolina and camped along the fertile Haw River. The river was used as a defensive barrier, from behind which Continentals launched patrols to test British defenses and gather provisions out of the reach of Commissary Stedman's foragers. Detached from the main army was Williams's light corps of roughly 700 troops, Lee's Partisan Legion, Pickens's Carolina militia regiment, and a battalion of Virginia riflemen.[31] Williams's careless defensive dispositions while his men were grinding corn at the mill enticed Cornwallis to send the 1,200 troops of Lieutenant Colonel Webster's reinforced brigade, the *23d* and *33d Regiments of Foot*, and *2d Battalion* of the *71st Regiment of Foot (Fraser's Highlanders)*, in a dawn attack on 6 March.

Tarleton's dragoons, screened by fog and darkness, easily breached the American picket line, forcing the surprised American units south of Reedy Fork Creek to run toward the fords. Reacting quickly, Williams planned to deploy his regulars on the high ground near Alamance Schoolhouse to screen the withdrawal of the main body.

---

[30] Roger Lamb, *An Original and Authentic Journal of Occurrences During the Late American War From its Commencement to the Year 1783* (Dublin, Ireland: Wilkinson and Courtney Dublin, 1809), 548.
[31] Conrad et al., *The Papers of General Nathanael Greene*, vol. 7, 408.

To cover the deployment of the Continentals, Pickens deployed skirmishers to slow Tarleton's advance with accurate rifle fire. Once in position around the schoolhouse, Williams's regulars fired multiple volleys, forcing the *33d Foot* and *Von Bose Regiment* to recoil from the creek's edge. While Webster regrouped, Pickens's skirmishers crossed and withdrew toward Greene's camp at High Rock Ford. Meanwhile, the *Guards Light Infantry* crossed upstream by the unguarded West Ford, although Webster's second attempt at the Horse Ford was again halted by disciplined volleys from the American rear guard. The stiff American resistance dissuaded Cornwallis from further pursuit, and Williams's command successfully broke contact to rejoin Greene's army. Neither side suffered more than a few casualties in a skirmish marked by leadership bungling. Cornwallis had fumbled an excellent chance to destroy many irreplaceable Continentals, while Pickens's militia regiment quit in disgust over Williams's perceived incompetence. The departure of Pickens prompted Greene to disband the flying army, with Williams later assuming command of the Maryland Continental brigade. The light infantry and Virginia rifles were attached to Lee's Legion and Washington's Partisan Regiment, forming corps of observations.[32]

Shortly thereafter, Greene received welcome reinforcements, Brigadier General John Butler's and Brigadier General Thomas Eaton's North Carolina militia brigades, roughly 1,500 fighters; Virginia State Line brigades commanded by Brigadier General Edward Stevens and Brigadier General Robert Lawson, totaling 1,200 troops; and the fresh but untested 2d Maryland Continental Regiment. Suitably reinforced, Greene marched toward the good defensive terrain of Guilford Courthouse, ready to offer battle.[33] Alarmed at his waning fighting strength, Cornwallis grasped at the opportunity to attack Greene early on 15 March 1781.[34]

Vignettes

In a letter written to Lieutenant Colonel Lee shortly before the Guilford Courthouse battle, General Greene outlined his tactical concept of the corps of observation, followed by some insightful comments on the operational situation:

> To Colonel Henry Lee, Jr. Camp at High Rock Ford [NC] March 9th 1781 9 oClock PM. Dear Sir. The light infantry is dissolved, and the Army will take upon itself an entire new formation. Col. Williams will join the line. . . . In lieu of the light Infantry two parties of observation, one to be commanded by you, and the other by Lt. Col. Washington. You are to attend to the enemies movements upon the left wing and Washington upon the right. It is my intention to give Col. Washington about 70 or 80 infantry and between three & four hundred riflemen to act with him. Col. Campbell . . . shall join you with about the same number of riflemen, and you and Col. Washington either separately or conjunctively as you may agree, to give the enemy all the annoyance in your power, and each to report to Head Quarters. . . . I am vexed to my soul with the Militia, they desert us by hundreds nay by thousands. I am now waiting for Gen'l [William] Caswell and the Continental troops to join us; which will not take place until tomorrow Night, nor can we organnise the Army sooner. After which we shall march in pursuit of the enemy as soon as possible, tho our force will not be very respectable. The intentions of the enemy is by no means fully explained. One thing is pretty certain, which is Lord Cornwallis don't wish a general action. If he did, he might have brought it on. It is quite uncertain whether the enemy will go by the way Cross Creek. I am rather inclind to believ he will go by the way of Colston, then

---

[32] Babits and Howard, *Long, Obstinate, and Bloody,* 47.

[33] "The Southern Campaign, 1781, from Guilford Court House to the Siege of York, Narrated in the Letters from Judge St. George Tucker to His Wife, the Southern Campaign as Narrated by St. George Tucker," in *The Magazine of American History*, vol. 7 (New York: A. S. Barnes, 1881), 39.

[34] Buchanan, *The Road to Guilford Courthouse*, 363–69.

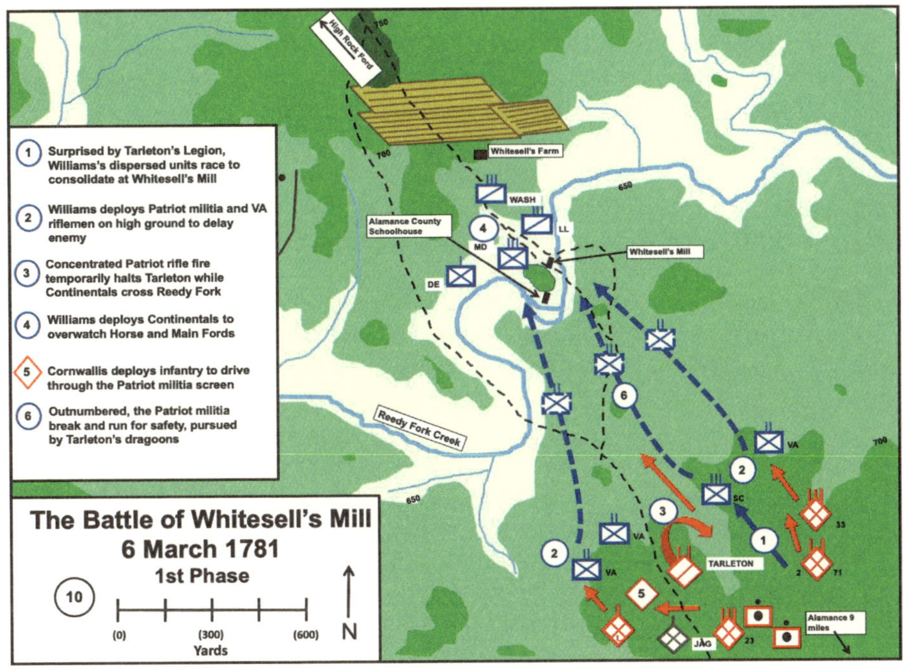

Map 10
Chuck Collins, Army University Press, and Harold Allen Skinner Jr.

Cross Creek. But let him march by what route he will, I beg you will observe his motions with great attention and give me the earliest information. I rely upon your exertions.[35]

Commissary Stedman's commentary on the operational situation follows:

Lord Cornwallis, having thus driven general Greene out of the province of North Carolina, returned by easy marches . . . to Hillsborough [North Carolina], where he erected the king's standard, and invited by proclamation all loyal subjects to . . . take an active part in assisting him to restore order and constitutional government. The loyalists in North Carolina were originally more numerous than in any of the other colonies: But the misfortunes consequent on premature risings had considerably thinned them. . . . Still, however, the zeal of some was not repressed; and. . . . considerable numbers were preparing to assemble, when general Greene, alarmed with the intelligence of their motions, and the presumed effect of lord Cornwallis's proclamation . . . took the resolution of again crossing the Dan and reentering North Carolina. . . . he had no intention of hazarding an action; but he foresaw that his return into the province would check the rising spirit among the loyalists; and he hoped. . . to interrupt their communications with the royal army. . . . A number of loyalists being ready to assemble, under a colonel Pyle. . . . Tarleton, with the cavalry, and a small body of infantry, was . . . to afford them assistance and protection. The American colonel Lee, having also received intelligence of the proposed insurrection, hastened . . . if possible, intercept and crush the loyalists before their junction with the British troops. The loyalists assembled, and on the twenty-fifth of February were proceeding to Tarleton's encamp-

---

[35] Conrad et al., *The Papers of General Nathanael Greene*, vol. 7, 415.

ment, unapprehensive of danger, when they were met in a lane by Lee, with his legion. The loyalists, unfortunately mistaking the American cavalry for Tarleton's dragoons, allowed themselves to be surrounded before they discovered their error. When at last it became manifest, they called out for quarter; but no quarter was granted; and between two hundred and three hundred of them were inhumanly butchered, while in the act of begging for mercy. Humanity shudders at the recital of so foul a massacre: But cold and unfeeling policy avows it as the most effectual means of intimidating the friends of royal government.[36]

Analysis

1. Perform a hasty course of action (COA) analysis for both Greene and Cornwallis: attack; defend; withdraw. What are the advantages, disadvantages, and risks to each COA?
2. Why were Cornwallis's efforts to raise Loyalist militia not as effective as the Americans'? What could Cornwallis have done to improve the situation?
3. Analyze Cornwallis's performance in light of his specified and implied mission. How does Cornwallis's focus on pursuit of Greene impact the big-picture campaign plan concerning the Carolinas?
4. What were the risks and consequences for the American campaign in the South if Greene lost? What were the potential gains for an American victory?
5. Consider the same questions from Cornwallis's perspective: what were the risks, consequences, and potential gains from battle with Greene's army?
6. Analyze Lieutenant Colonel Lee's decision-making at the skirmish with Pyle's Loyalists. What were some other options or courses of action he could have taken? What of the perceptions of a massacre; did it help or hurt the American cause?
7. What are some ways Greene and Williams could have better handled the militia's response after the Whitesell's Mill skirmish?

Stand 6: The Skirmish at New Garden Meeting House

Directions

From Whitesell's Mill, return to NC-16, and turn right (south). Drive 1.4 miles (2.3 km) and turn right on Huffine Mill Road. Drive about 12 miles (19.3 km) on Huffine Mill Road to the intersection of US 70/220 (West Wendover Avenue). Turn right (west) on West Wendover Avenue and drive 4.9 miles (7.9 km) to Green Valley Road. Turn left (south) and take an immediate right (northwest) on Friendly Avenue. Drive 3.5 miles (5.6 km) to the intersection of Friendly Avenue and College Road. Park in the large shopping center parking lot to the south.

Orientation

This location overlooks where Lieutenant Colonel Banastre Tarleton's *British Legion* dragoons, and Lieutenant Colonel Henry Lee's Corps of Observation fought the first skirmish of the Guilford Courthouse battle on 15 March. About 220 meters to the northeast is the New Garden Friends Meeting House, which in 1781 was a small cluster of buildings that housed a congregation of the Religious Society of Friends. About 1.8 miles (3 km) to the northeast, near the intersection of the modern New Garden Road and Hobbs Road, is "The Lane" where the opening skirmish took place around 0700. The intersection of the West Friendly Avenue and New Garden Road, about 220 meters to the

---

[36] Stedman, *The History of the Origin, Progress, and Termination of the American War*, vol. 2, 332–34.

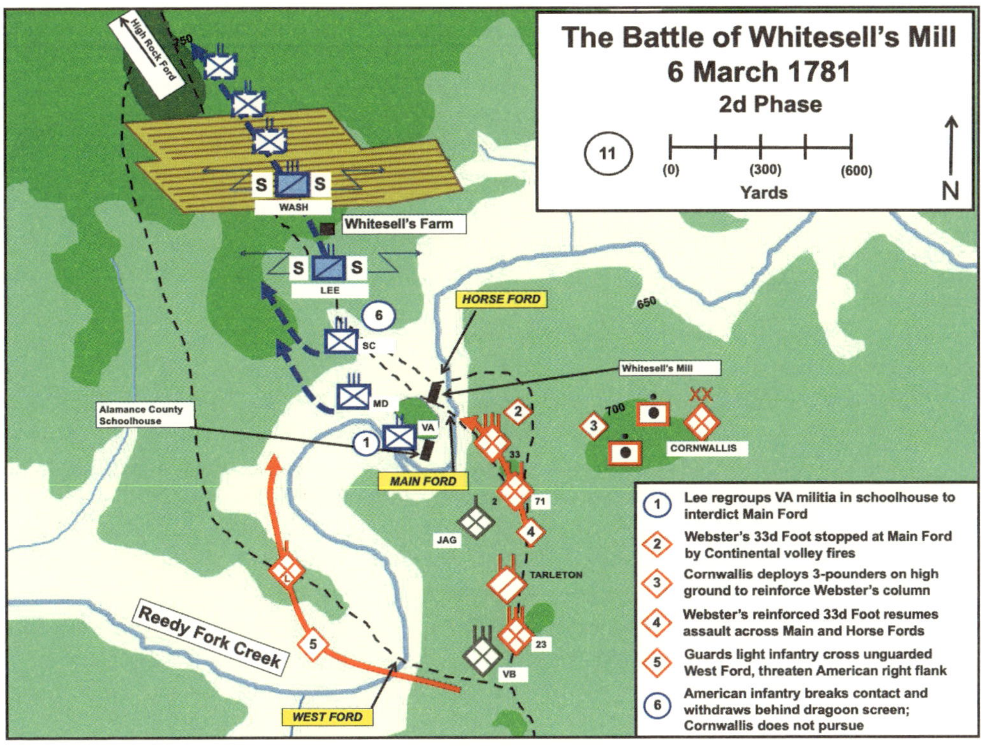

Map 11
Chuck Collins, Army University Press, and Harold Allen Skinner Jr.

northeast, marks the approximate location for a second skirmish that took place around 0900. About 0.6 miles (1 km) north of this location is the intersection of Ballenger Road with the Great Salisbury Road, the site of a third skirmish that took place around 1000. To the northeast about 4.5 miles (7.2 km), is the location of Guilford Courthouse. Eight miles (12.8 km) to the southwest was the Deep River Friends Meeting House, where Cornwallis's logistics trains were positioned during the battle.

Description

Established in 1751 by the pacifist Religious Society of Friends, commonly known as Quakers, New Garden had about 120 houses clustered on a 50-acre plot of rich farmland. With the notable exception of Nathanael Greene, the Quakers largely refused to bear arms, earning the scorn of both Loyalists and Patriots. However, both sides were happy to collect provisions from the productive Quaker farms and deposit sick and wounded men for treatment.

Early on 14 March, Greene sent Lee's Corps of Observation to New Garden to screen the Great Salisbury Road, Cornwallis's likely avenue of approach from the Deep River Friends Meeting House, about 8 miles (12.8 km) to the southwest. Lee's corps included the Partisan Legion, about 150 mounted dragoons and light infantry, a battalion of Virginia light infantry, and a company of Continental light infantry. From the meeting house, Lee detached Lieutenant James Heard's platoon of dragoons, to conduct area reconnaissance near Cornwallis's camp. Around 0200 on the morning of 15 March, Heard sent the first of a series of messages back to Lee reporting on movement within the British camp. Around 0400, Lee alerted Greene of the probable movement of the British Army, then ordered his troops to wake up and take a hasty breakfast while waiting for orders.

Greene responded quickly, ordering Lee to perform a reconnaissance in force toward the British camp. Lee's forces mobilized rapidly with dragoons in the lead and the slower infantry trailing in support. Soon thereafter, Heard's dragoons fired a volley at the advanced guard of Tarleton's *Legion*, comprising 272 mounted dragoons supported by 100 troops of the *Guards Light Infantry* and a company of 84 rifle-armed Ansbach jägers. The poorly aimed volley briefly checked Tarleton's advance along the Great Salisbury Road. Tarleton responded to Heard's volley with an immediate attack. After passing the New Garden Meeting House, Tarleton's vanguard collided with a troop of Lee's dragoons deployed in a hasty ambush with their flanks protected by two long fences. A brief clash of sabers ended with several Loyalist dragoons unhorsed and captured, while the remainder swiftly withdrew along a dirt trail to the east of the Salisbury Road. Responding to the sounds of battle, Lee and several of his troopers blundered into a hasty ambush by the *Guards Light Infantry*. By the time Lee had remounted on a borrowed horse, his supporting light infantry and riflemen had moved up in support. A sizable meeting engagement developed as Tarleton's outnumbered light infantry were reinforced by the lead battalion of the *23d Regiment of Foot*. With the arrival of the *23d Foot*, Lee shifted to a delaying defense, skillfully employing his dragoons in the canalizing terrain to frustrate Tarleton's attempted pursuit, thus allowing the American infantry time to reorganize at the Ballenger and Great Salisbury "Cross Roads." Lee then consolidated his dragoons with the infantry and continued a delaying action along the Great Salisbury Road until 1000, when Lee broke contact and withdrew, unpursued, to Guilford Courthouse. Both sides suffered approximately 40 casualties in this opening engagement, but Lee's corps clearly dominated the fight and kept Tarleton from gathering actionable intelligence about Greene's dispositions. By contrast, Greene received regular reports from Lee's couriers, and Americans were given time to perfect their troop deployments at Guilford Courthouse.

Vignettes

Lee's third-person account of the New Garden Meeting House engagements.

> Lee . . . immediately advanced on the road toward the Quaker meeting-house, with orders to post himself [there] . . . Lieutenant Heard, of the Legion cavalry, was detached in the evening with a party of dragoons to place himself near the British camp, and to report from time to time such occurrences as might happen. About two in the morning [Lt. Heard] communicated that a large body of horse were approaching the meeting house, which was not more than six miles from our head-quarters, and near the point where the road from Deep River intersects the great road leading from Salisbury to Virginia. The intelligence received was instantly forwarded to the general, and Heard was directed to . . . discover whether the British army was in motion. . . . And at length he communicated his various attempts to pass down the flank . . . had proved abortive, having been uniformly interrupted by patrols ranging far from the line of march; yet he . . . heard the rumbling of wheels, which indicated a general movement. . . . Lee was directed to advance with his cavalry . . . and to ascertain the truth. Expecting battle . . . the van was called to arms at four in the morning, and to take breakfast with all practicable haste. This had just been finished, when the . . . order from the general was communicated. Lieutenant-Colonel Lee instantly mounted, and took the road to the enemy . . . having directed the infantry and rifle militia to follow. . . . The cavalry had not proceeded above two miles, when Lee was met by Lieutenant Heard, followed leisurely by the enemy's horse. Wishing to approach nearer to Greene, and . . . gain the proximity of the rifle militia and Legion infantry, lest the British army might be up, as was suspected, Lee ordered the column to retire by troops. . . . The rear troop, under Rudolph, going off in full gallop . . . the British commandant flattered

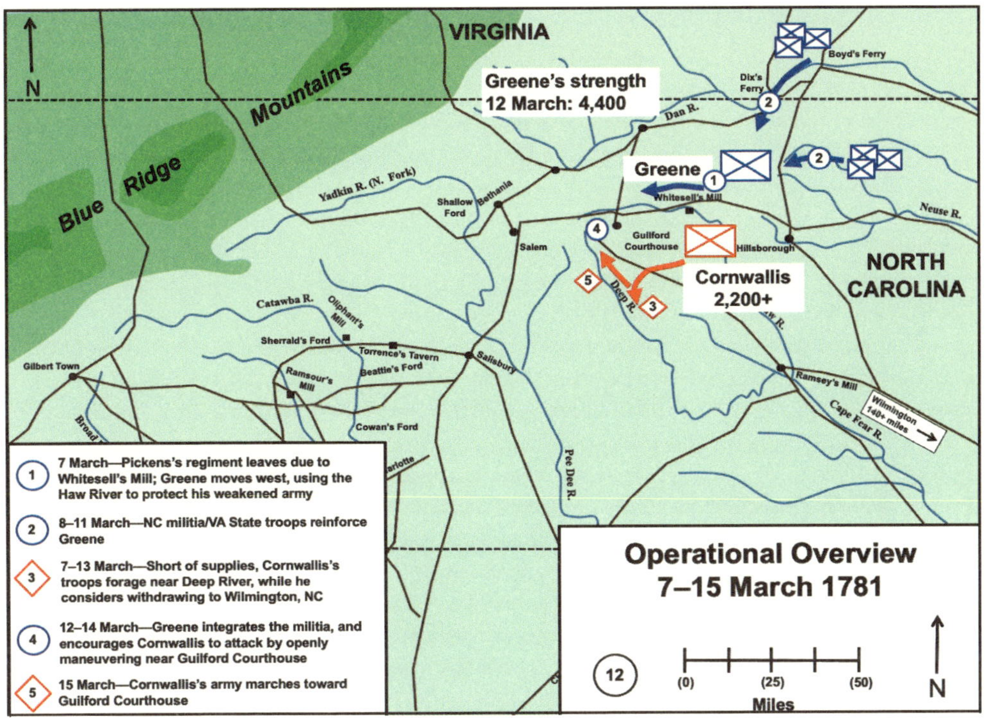

Map 12
Chuck Collins, Army University Press, and Harold Allen Skinner Jr.

himself with converting this retrograde movement into rout. . . . the enemy emptied their pistols, and then raising a shout, pushed a second time. . . . At this moment, Lee ordering a charge, the dragoons came instantly to the right about, and, in close column, rushed upon the foe. This meeting happened in a long lane, with very high curved fences . . . which admitted but one section in front . . . . Tarleton sounding a retreat, the moment he discovered the column in charge. The whole of the enemy's section was dismounted. . . . Tarleton retired with celerity . . . took an obscure way . . . toward the British camp—while Lee . . . followed the common route by the Quaker meeting-house, with a view to sever the British lieutenant-colonel from his army. . . . By endeavoring to take the whole detachment, he [Lee] permitted the whole to escape. . . . As Lee, with his column in full speed, got up to the meeting-house, the British guards had just reached it; and . . . gave the Americans a close and general fire. The sun had just risen above the trees, and shining bright, reflection from the British muskets . . . frightened Lee's horse so as to compel him to throw himself off. Instantly remounting another, he ordered a retreat . . . . while the cavalry were retiring, the Legion infantry came running up with trailed arms, and opened a well-aimed fire upon the guards, which was followed in a few minutes by a volley from the riflemen under Colonel Campbell. . . . The action became very sharp, and was bravely maintained on both sides. . . . Lee being convinced, from the appearance of the guards, that Cornwallis was not far in the rear, drew off his infantry; and covering them . . . retired towards the American army. General Greene, being immediately advised of what had passed prepared for battle.[37]

---

[37] Henry Lee, *Memoirs of the War in the Southern Department of the United States* (New York: University Publishing, 1869), 272–75.

Lieutenant Colonel Tarleton's version of the New Garden skirmishes:

> On the 14th of March, his lordship determined to advance upon the Americans at Guildford, and bring on an engagement, that he thought they would not avoid. . . . The main body at daybreak marched toward the enemy's camp. The cavalry, the light infantry of the guards, and the yagers, composed the advance guard. . . . The British had proceeded seven miles on the great Salisbury road to Guildford, when the light troops drove in a picket of the enemy. A sharp conflict ensued between the advanced parties of the two armies. In the onset, the fire of the Americans was heavy, and the charge of their cavalry was spirited . . . the gallantry of the light infantry of the guards, assisted by the legion . . . [and the arrival of] the 23rd regiment . . . [forced Lee's dragoons to retreat] with precipitation along the main road. . . . The pursuit was not pushed very far, as there were many proofs . . . that General Greene was at hand. . . . An engagement was now become inevitable.[38]

Analysis

1. Lee's candid admission of his failure to trap Tarleton's force leads to some interesting questions:
   a. What was Lee's stated mission?
   b. What of his implied mission?
   c. How did Lee's actions at the Friends Meeting House meet or deviate from Greene's specified and implied commander's guidance?
2. Both sides used a complementary mix of dragoons, riflemen, and heavy infantry. What are some modern insights that we can glean from such examples?

DAY TWO: TACTICAL LEVEL GUILFORD COURTHOUSE STANDS
Stand 1: The Road to Guilford Courthouse (Operational Orientation)
Directions

The suggested starting point for day two is the Guilford Courthouse National Military Park (GCNMP). The Battlefield Visitors Center is administered by the National Park Service (NPS) and is located at 2332 New Garden Road, Greenville, North Carolina, 27410. Viewing of the 30-minute-long official NPS video and 10-minute animated battle map is suggested to refresh memories before the terrain walk. The large concrete pad to the east of the visitor's center, away from foot and vehicle traffic, is used for the first outside orientation.

Orientation

Guilford Courthouse is located approximately 100 miles (161 km) northeast of Charlotte, North Carolina. The two major British bases in South Carolina sat well to the south. Ninety Six was approximately 240 miles (386 km) to the southwest; Camden was almost due south at 160 miles (257.5 km). By contrast, the key American logistics base at Danville, Virginia, was only 45 miles (72.4 km) to the northeast.

Upon arriving at this location, the instructor should first orient the students to the cardinal directions of the compass. Next, point northeast toward Guilford Courthouse village, about seven-tenths

---

[38] LtCol Banastre Tarleton, *A History of the Campaigns of 1780 and 1781 in the Southern Provinces of North America* (London: T. Cadell, 1787), 270–71.

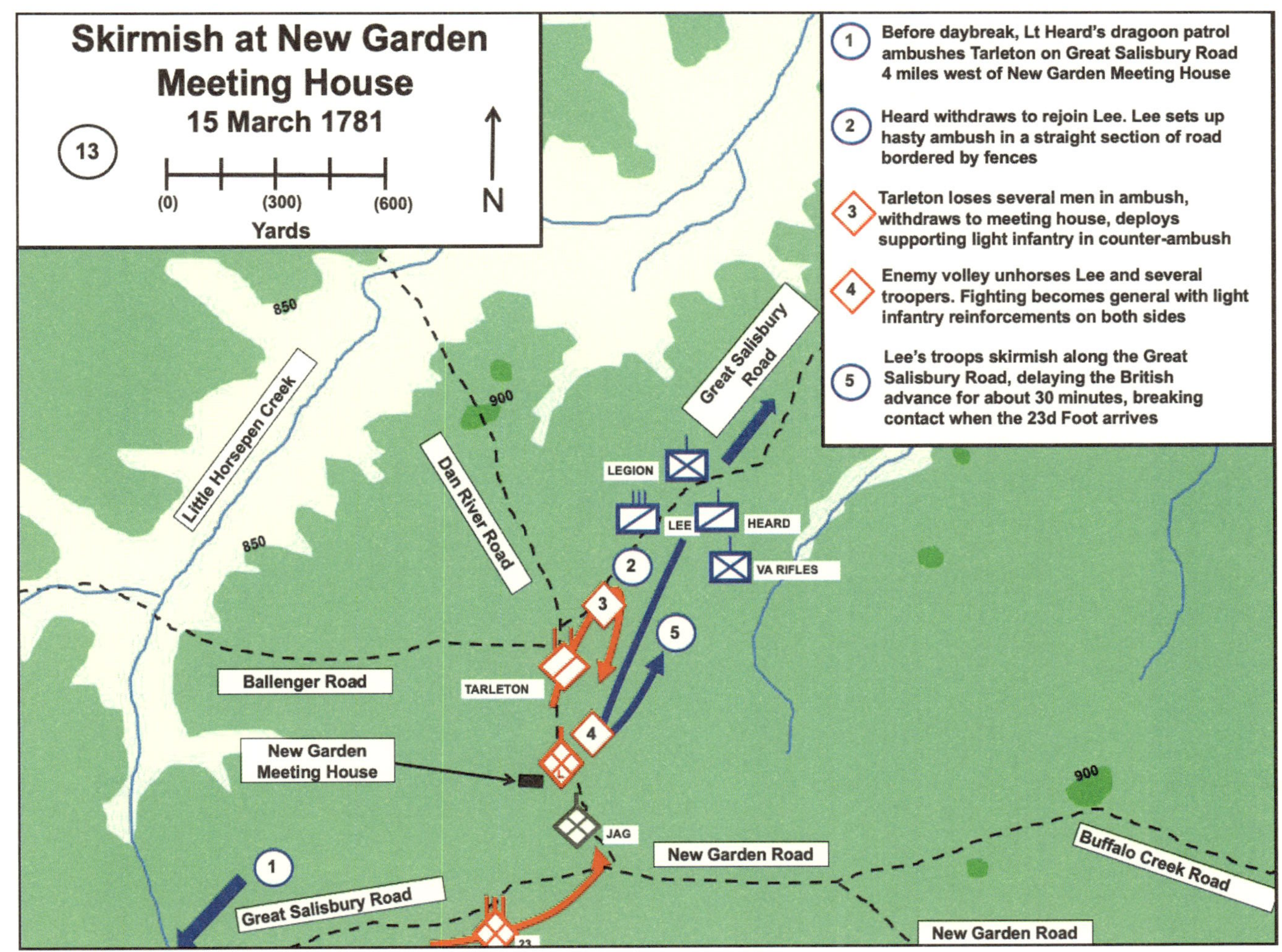

Map 13
Chuck Collins, Army University Press, and Harold Allen Skinner Jr.

of a mile (1.05 km) from here. About 50 meters to the north of the visitor's center is the trace of the historic New Garden Road that ran between New Garden Meeting House and Guilford Courthouse. The road outside the park is paved, while the path within the park is gravel and sand, probably similar to that of 1781. Compared to today, the wooded areas of 1781 would have consisted mostly of mature trees, with little undergrowth between, as free ranging cattle would have closely cropped the vegetation. As the battle took place in March, the open fields would have been either muddy from spring plowing or covered with stubble from the previous harvest, while the deciduous trees would have had only immature leaves. The weather in the days leading up to the battle was probably mild, with alternating cool and warm days. Enough rain fell during this period that the open fields were damp and muddy, but not so much as to significantly inhibit cross-country movements.

Description

Frustrated with his failure to corner the Continental Army before it could escape to its Virginia sanctuary, Cornwallis withdrew to Hillsborough, North Carolina, where on 22 February 1781 he "erected the King's Standard."[39] While his hungry redcoats foraged for scarce provisions and attempted to enroll new Tory recruits, Cornwallis sought for the means to lure Green into battle. While his regiments rested and refitted, Greene badgered the governors of Virginia and North Carolina for supplies and reinforcements. Time was essential, as Greene could not afford to leave Cornwallis undisturbed to regain his strength. On 19 February 1781, Williams's light infantry was sent back into North Carolina, shortly followed by the balance of the American Army. After several weeks of cat-and-mouse play between the two armies, punctuated by several minor skirmishes and the major skirmish at Whitesell's Mill, General Greene marched his army to Guilford Courthouse to offer battle to the British Army.

Vignettes

Cornwallis's hopes for a quiet winter to rest and recruit new Tory militia at Hillsborough were upset by Greene's unexpected return to North Carolina.

> Where he [Cornwallis] erected the king's standard, and invited by proclamation all loyal subjects to repair to it, and take an active part in assisting him to restore order and constitutional government. . . . But the misfortunes consequent on premature risings had considerably thinned [the Tories]. . . . And those who remained were become cautious from the recollection of past miscarriages. . . . [however] considerable numbers were preparing to assemble, when general Greene, alarmed with the intelligence of their motions, and the presumed effect of lord Cornwallis's proclamation . . . took the resolution . . . re-entering North Carolina. Even with this addition to his numbers [the Virginia militia], he had no intention of hazarding an action; but he foresaw that his return into the province would check the rising spirit among the loyalists.[40]

Writing to North Carolina governor Abner Nash, General Greene gave a good executive summary and analysis of the operational situation in late February 1781.

> To Governor Abner Nash of North Carolina, Head Quarters, Boyds Mill [NC], March the 6th 1781.
>
> > Dear Sir, I have to acknowledge the receipt of your Letter dated the 23d of February. . . . Compelled by a superior force I was obliged to submit to the painful necessity of retreating to Halifax Court House in Virginia, & there wait for reinforcements.

---

[39] Stedman, *The History of the Origin, Progress, and Termination of the American War*, vol. 2, 332.
[40] Stedman, *The History of the Origin, Progress, and Termination of the American War*, vol. 2, 332–33.

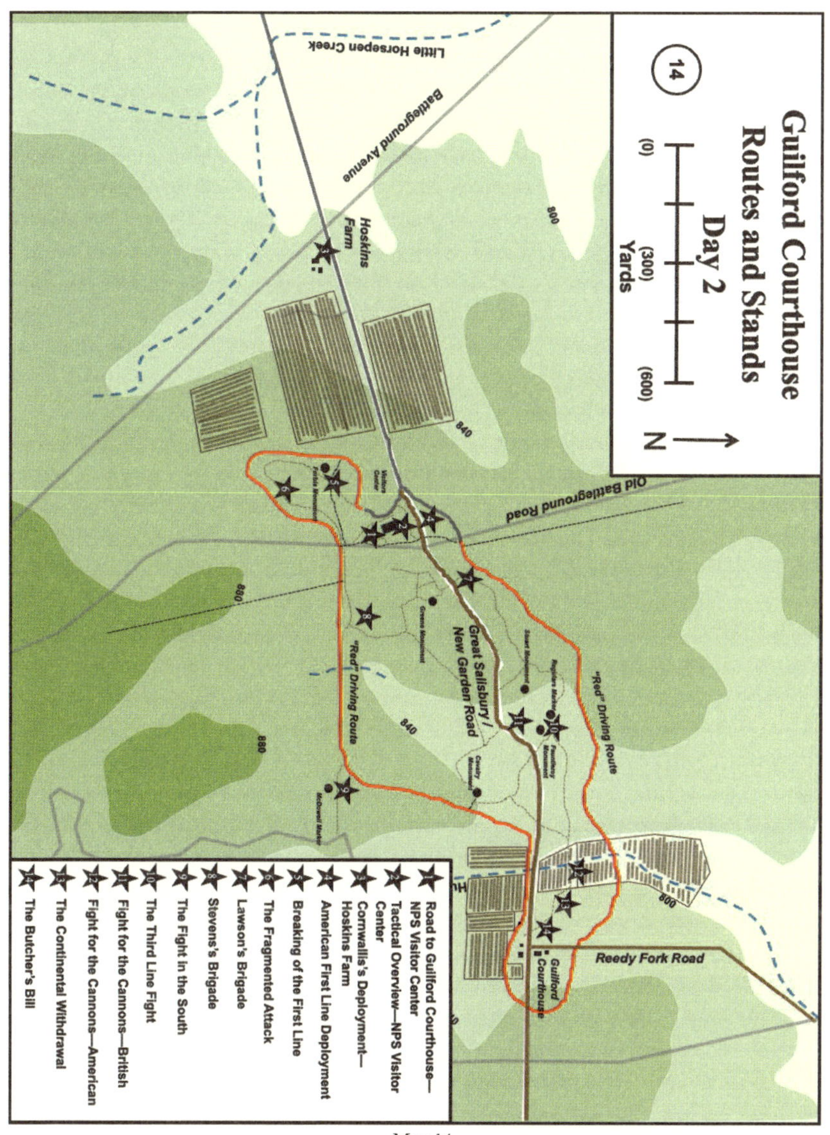

Map 14
Chuck Collins, Army University Press, and Harold Allen Skinner Jr.

. . . The Enemy during my stay gave evident proofs of a design to push though Virginia . . . and the People in the adjacent Counties flew to Arms universally, but as soon as the Enemy filed off towards Hillsborough that spirit of opposition in a great measure failed & the reinforcements which I had collected were falling off every day. My force was upon such a loose and uncertain footing, that no regular plan of operation could be adopted or any decisive measures taken to recover the Country from the calamities of wanton depredations. Some partial efforts have been made with success, & the best of consequences have resulted from them. The Tories in several instances have been baffled & our Light parties have taken a number of Prisoners from time to time. The enemy at present lays . . . within 15 miles [24 km] of our Army, & is encamped in so compact a Body that it is difficult to attack them. . . . It is particularly painful to me to be obliged to prosecute a War under the disadvantages of depending on Militia and short enlistments. Every moment hazards the reputation of any Man who depends upon them. If it is possible to raise your regular Troops, it had better be done. Every effort ought to be made to effect it, and if it shall be found impracticable to compleat it fully, let it be done in part. . . . My dear sir the present crisis is important, & requires the most decicive exertions. From the state of things I can promise you no security or positive grounds of assurance [for] the safety of your Country or property. The desertions which prevail in the Militia will, I am afraid Dwindle my force into a body too weak to act offencively. They get tired out with difficulties, and for want of discipline bid defiance to authority. Every effort to save a Country under such circumstances must prove ineffectual. Nothing can or will do but a regular Army, and nothing ought to be left undone that can possibly effect the raising of one.[41]

Analysis

1. What was the comparative strategic situation in 1781?
2. What was the comparative operational plan for the American and British sides?
3. How did Cornwallis's operational plan meet or deviate from Clinton's strategic vision?
4. What of Greene's operational plan compared to General Washington's strategic vision?
5. What contemporary comparisons/contrasts can we draw about Greene's Fabian strategy of exhaustion employed against the British?[42]

Stand 2: The Tactical Situation

Directions

From stand 1, the east side of the visitor's center, move north approximately 100 meters to the intersection with a gravel path running east and west, which is the historical trace of the New Garden Road.

Orientation

This location overlooks the avenue of approach used by Cornwallis's army during the morning of 15 March 1781 as it marched up from the southwest along the Salisbury Road (today's New Garden Road), preceded by Tarleton's dragoons. Facing north, the village of Guilford Courthouse is about 900 meters to the northeast. About 500 meters to the southwest was the Hoskins farmstead, where Cornwallis deployed his army for battle. The location of the American first line was about 100 meters to the west. In between was a patchwork of plowed fields and small copses of trees. To the east of this position, running from northeast to southeast (across the modern-day Old Battleground Road), was a thick belt of timber with trees scattered across the uncultivated spaces.

Description

Screened by detachments from Lieutenant Colonel Lee's Corps of Observation, the well-rested American Army at Guilford Courthouse had time to leisurely eat breakfast and deploy into their respective battle lines. Even though Greene had previously done a general survey of the courthouse

---

[41] Conrad et al., *The Papers of General Nathanael Greene*, vol. 7, 401–2.

[42] George Washington to John Hancock, 8 September 1776, Founders Online, National Archives and Records Administration, accessed 22 June 2020. The term *Fabian strategy* derives from the delaying strategy employed by Roman general Quintus Fabius Maximus Verrucosus against the superior Carthaginian army of Hannibal. Washington and Greene learned to avoid major battle in favor of attacks against isolated British detachments and raids to seize provisions from British possession—all designed to weaken the enemy's will to continue.

area, the morning of 14 March was spent performing a reconnaissance of the intended battlefield and drilling his units and officers on their intended actions. Despite the advance warning of the British approach, Greene did not order the construction of obstacles or entrenchments. Consequently, the only protection on the American first line was a flimsy split rail fence, while trees and rolling terrain provided better cover and concealment for the second and third battle lines.

In comparison to the well-fed and rested Americans, the British officers and soldiers felt the cumulative effects of operating on half rations and little sleep for several weeks. Even before the day of the battle, Cornwallis's troops had been on the road, marching up from the Speedwell Furnace area (near Rockingham, North Carolina) since 0300 the previous morning. By this point, Cornwallis urgently needed to fight Greene before the combat readiness of his army fell to unacceptable levels from sickness and desertion.

After intermittent mild and rainy weather, the morning of 15 March dawned clear. The average high temperature in March for this area is around 60 degrees, with accounts of the battle describing a slightly warmer than average late-winter afternoon, without wind sufficient to clear clouds of smoke generated by muskets. The battle started around 1300, so late that afternoon shadows would have gradually accumulated in the woods screening the American second and third lines, augmenting the concealment offered by the thick gunpowder smoke.

Vignettes

Greene's assessment of Cornwallis's precarious situation before the battle:

> Immediately on his arrival at Hillsborough, appeared the British commander's celebrated proclamation . . . calling upon his loyal and faithful subjects to repair without loss of time to his standard, equiped with arms and furnished with ten days provisions. . . . to be formed into regiments. . . . But, suddenly the crowd began to drop away, and the dissipation of his proud hopes was soon accounted for by intelligence, that Greene, being reinforced, had re-crossed the Dan. . . . It was on this occasion that Lord Cornwallis wrote to the British ministry that he was surrounded by timid friends and inveterate enemies. The dangers of the situation began now to open on his view; and to fight his antagonist, became . . . a matter of necessity, as it had been before a matter of choice. The waning state of his stores—the falling off of his men by desertion—the defection of the wary loyalists—the alarm from all quarters, of the advance of the hostile militia, and not less, a consciousness that he had made no friends in his advance through the country—plainly showed that he had now no alternative, but conquest or destruction. The army of Greene must be dissipated or driven off, or his own discomfiture was scarcely avoidable.[43]

Commissary Stedman recorded the British perspective:

> As lord Cornwallis retired [from Hillsborough] the American army advanced; and general Greene . . . took post between Troublesome Creek and Reedy Fork; but not thinking himself yet strong enough to risque an action, he changed his position every night in order to avoid the possibility of it. . . . [After the Whitesell's Mill skirmish on 6 March 1781] Greene . . . retreated over the Haw, in order to preserve his communications with the roads by which he expected his supplies and reinforcements. . . . General Greene, thus powerfully reinforced, knowing that the time of service of the militia would soon expire, determined to avail himself of his present strength by offering battle to lord Cornwallis. Accordingly . . . moved forward to Guildford Court-house, within twelve miles [19 km] of the British

---

[43] Johnson, *Sketches of the Life and Correspondence of Nathanael Greene, Major General of the Armies of the United States, in the War of the Revolution*, vol. 1, 447–48.

army, which, since his last retreat, had taken a new position at the Quakers' Meeting-house in the Forks of the Deep River. The near approach of general Greene . . . since he was joined by his reinforcements, indicated an intention of no longer avoiding action, lord Cornwallis embraced with much satisfaction the proffered opportunity of giving him battle.[44]

Cornwallis's operational order for the assault on Guilford Courthouse:

Genl. After Orders, 14th March 81

The army to be under arms & the bat[talion] Horses Loaded ready to march precisely At half past five O'Clock tomorrow Morning In two Columns in the following Order:

Left Column	Right Column
Advanced Gd Commd by	An Officer & 12 Dragoons
Lt C. Tarleton	Lt. Col. Hamilton's Regt.
Jagers	
Lt Infantry Guards	
Cavalry	
Lt. Col. Webster's Brigade	
Regt. De Bose.	
Brig. Of Guards	
Guns as Usual	

A Detachment of two Captns., three Sub[alterns] & one Hundred Men from the Regts. Which form the left Column to be Composed of Serviceable men but not the best marchers.

Detail for the Detachment

	Capt.	Sub.	S:	C:	P:
Brig. of Guards	1		2	2	35
Lt. C. Webster's Brig.	1	2	3	3	45
Regt. De Bose	1	1	1	20	
Total	2	3	6	6	100

The troops to send for meal at ½ past 4 in the Morning The Baggage will move With the right Column.

B. Orders

The detachment To be form'd (as Specify'd by Genl. Orders) from the whole of the Brige. Captn. [Charles] Horneck, the Officer for this Duty.[45]

A contemporary description of the American tactical dispositions at Guilford Courthouse sheds some light on General Greene's purpose, key tasks, and envisioned end state.

---

[44] Stedman, *The History of the Origin, Progress, and Termination of the American War*, vol. 2, 336–37.

[45] A. R. Newsome, ed., "A British Orderly Book, 1780–81, Part IV," *North Carolina Historical Review* 9, no. 4 (July 1932): 387.

If the reader will now run his eye over the whole field of battle, he will be struck with observing that the army is all in lines—that there is no corps set apart as a reserve—and this will lead him to penetrate the views which directed the whole arrangement, and governed many of the events of the battle. The regular troops are so stationed, as to serve as a reserve to the whole army. There were two avenues of retreat. . . . The continental troops are drawn with a view to secure the choice of these routes, according to the point to which the enemy should direct his attack. If pressed upon the right wing, the avenue by the left could be resorted to; if pressed on the left that by the right would answer the object in view. The resolution which governed every movement of the American general was, in no event, to hazard the destruction of the regular troops. To cripple the enemy by his militia and light troops, and insure their retreat under protection of his regulars, was his motive. If, in pursuit of these objects, fortune should prove propitious, there was ample preparation made for availing himself of the event. The cavalry and rifle corps, besides covering the wings of his own line, would be at hand to strike at the exposed flanks of a retreating enemy.[46]

Analysis

1. What risks did General Greene assume with accepting battle at Guilford Courthouse? What measures could he have taken to mitigate the risk?
2. Knowing his logistics situation, did Cornwallis make a sound decision with the information at hand to attack the American Army? What were his alternatives? What measures could he have taken to mitigate the risk of attacking?
3. What were some mental factors or perceptions that could have influenced Greene's or Cornwallis's decision-making? What similar factors influence decision-making today?
4. What processes did Green and Cornwallis use to obtain and analyze information to develop an accurate picture of the enemy and battlefield? How does this process compare to today's intelligence processes?
5. Using *Offense and Defense*, ADRP 3-90, chapter four, or *Commander's Tactical Handbook*, MCRP 3-30.7, part one, "Defensive Operations," as a guide, what were some possible defensive planning shortfalls by Greene?[47]

Stand 3: Cornwallis Deploys for Battle—Hoskins Farmstead
Directions

From stand 2, retrace your steps to the parking lot to move by vehicle to stand 3, which is located at the Hoskins Farm site on New Garden Road, about 500 meters west of the visitor's center. The use of vehicles to move to the site is recommended for safety reasons, as there is no public easement across the intervening private property and the heavily trafficked New Garden Road. The Hoskins Farm has 30 parking spaces sufficient for cars and vans, but advanced coordination is needed to use bus transportation to this site. After arriving at the parking lot of the Hoskins Farm, walk north about 100 meters, following the concrete walk to the "British Attack" interpretive panel overlooking the New Garden Road.

---

[46] Johnson, *Sketches of the Life and Correspondence of Nathanael Greene, Major General of the Armies of the United States, in the War of the Revolution*, vol. 2, 7–8.
[47] *Offense and Defense*, ADRP 3-90 (Washington, DC: Headquarters U.S. Army, 2012), 4-6–4-24; and *Commander's Tactical Handbook*, Marine Corps Reference Publication (MCRP) 3-30.7 (formerly MCRP 3-11.1A) (Washington, DC: Headquarters Marine Corps, 2016), 43–58. The author identified the following: mission command, exploit the terrain, security, "mass the effects of combat power," plus subtasks within the intelligence and protection domains.

Orientation

This location overlooks the approximate point where the British marching columns deployed into combat formation. During the walk to the interpretive panel, point out the slight but noticeable rise in the terrain from west to east. After reading the panel, move forward to the wooden rail fence paralleling the road. Face left (southwest) along New Garden Road to direct the students' attention to the British deployment point just across the bridge crossing Horsepen Creek, which is roughly 1.2 miles (2 km) from this location. Horsepen Creek today is trapped in the urban sprawl of Greensboro, thus unsuitable as a staff ride stand. To the right about 100 meters was where the American first line crossed the New Garden Road.

Description

In 1781, most of the ground to the east of the Hoskins House was a checkerboard of fallow farmland or "oldfields." Individual field plots were subdivided by split rail fence, and a large rail fence ran along the east edge of the fields, the location where Greene elected to deploy his North Carolina militia. On the day of the battle, line of sight was obstructed only by split rail wood fences, while clear skies meant good visibility to the naked eye. The Continental artillery section in the first line fired directly down this road to the west, attempting to enfilade the British column as the battalions deployed into line formation. The slight rise from west to east put the British troops at a tactical disadvantage, as musket-armed troops tended to aim high when firing uphill.

The lead elements of Cornwallis's corps arrived at this general location around 1230 on 15 March 1781. Warned by his scouts about the American militia line, Cornwallis ordered his troops to deploy into combat formation as soon as the battalions crossed the Horsepen Creek. To provide covering fire during the transition, Cornwallis ordered the *Royal Artillery* section to emplace 3-pounders along the road near the Hoskins farm. The well-trained British gun crews quickly unlimbered and began an inaccurate exchange of fire with the Continental 6-pounder guns to the east. Few casualties were reported from the artillery duel, although a solid shot struck down Lieutenant Augustus O'Hara, a younger relation to Brigadier General Charles O'Hara. Greater than the physical effect of the fire was the moral effect of the gunfire on the inexperienced North Carolina militiamen. The British deployment took roughly 30 minutes, starting on the left with Lieutenant Colonel Webster's brigade, comprising the *23d Regiment of Foot (Royal Welch Fusiliers)* and *33d Regiment of Foot*. In support of Webster's regiments was the *2d Guards* of O'Hara's *Brigade of Guards*. Major General Alexander Leslie's brigade with the *71st Foot (Highlanders)* and the German *Von Bose Regiment* deployed on line to the British right, with the *1st Guards* echeloned behind in support. Cornwallis remained on the road, where he retained control of the 3-pounders, and the *Guards Light Infantry* and grenadier companies, jägers, and Tarleton's *Legion* as a reserve. Cornwallis did not pause for reconnaissance or to send out skirmishers to develop the American line, pausing only to consult with his Loyalist guides about the lay of the land. Once the battle line was formed around 1300 hours, Cornwallis ordered his regiments forward in a frontal attack against the visible enemy militia line.

Vignettes

In his after action report to Lord George Germain, Cornwallis described the battlefield:

> Immediately between the head of the column and the enemy's line was a considerable plantation [the Hoskins farmstead], one large field of which was on our left of the road, and two others, with a wood of about 200 yards broad between them, on our right of it; beyond these fields the wood continued for several miles to our right. The wood beyond the plantation in our front, in the skirt of which the enemy's first line was formed, was about

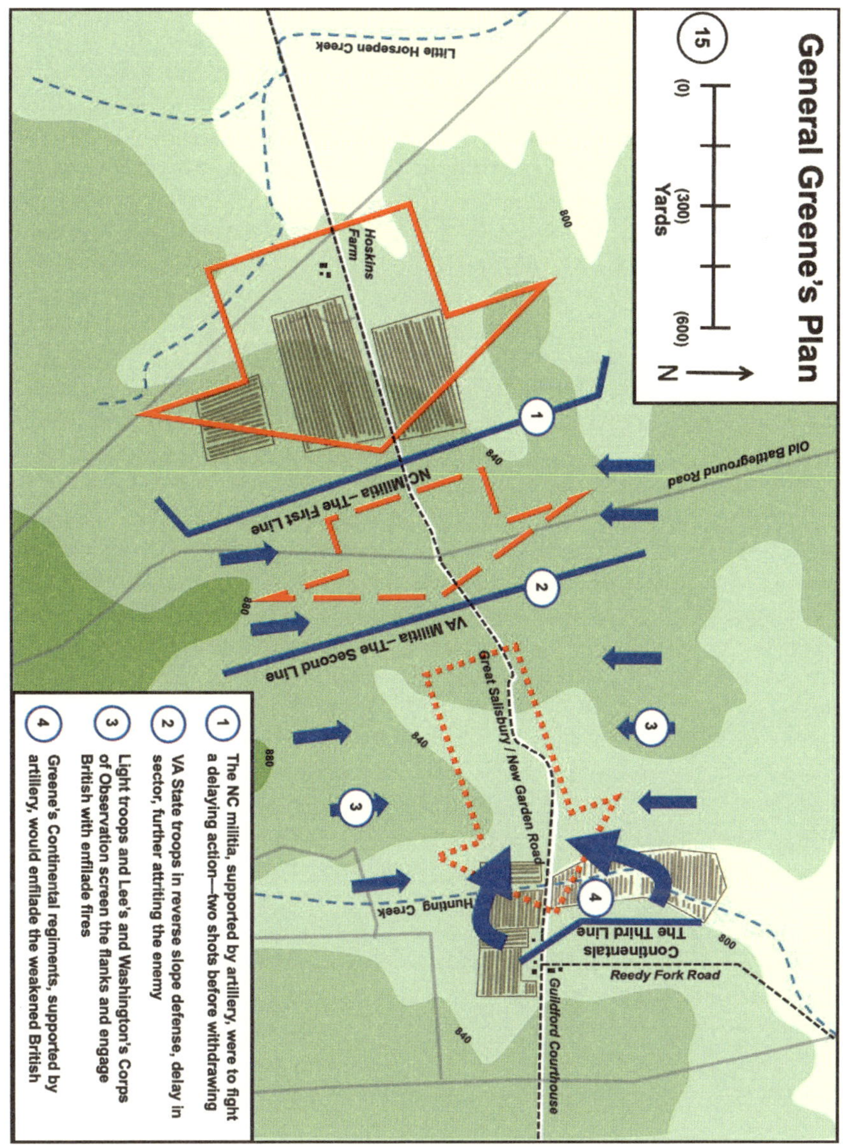

Map 15
Chuck Collins, Army University Press, and Harold Allen Skinner Jr.

a mile in depth, the road then leading into an extensive space of cleared ground about Guilford Court-house.[48]

Lieutenant Colonel Tarleton's description of the British deployment and movement to contact:

As the front of the British column approached the open ground facing the American position, the enemy's six pounders opened from the road, and were immediately answered by

---

[48] Cornwallis to Lord George Germain, 17 March 1781, "Key Original Source Outline for the Battle of Guilford Courthouse," working files, Guilford Courthouse National Military Park, no date, hereafter Cornwallis letter, 17 March 1781.

<table>
<tr><td>

Weather:
- Light rain in days leading up to battle make fields and unimproved roads muddy and slick
- 15 March 1781
    - Day dawns clear and humid
    - Rain returns in the late afternoon
    - Rain and damp increase risk of misfire with flintlock weapons
- Light:
    - Beginning Morning Nautical Twilight (BMNT): 0555
    - End Morning Nautical Twilight (EMNT): 0625
    - Beginning Evening Nautical Twilight (BENT): 2012
    - End Evening Nautical Twilight (EENT): 2041
    - Lee's troopers affected by sunlight reflecting off British armor
    - British assault begins around 1300 = about 6 hours of usable daylight

</td><td>

Observation:
- Limited by undulating terrain and thick woods at second and third lines
- Otherwise unobstructed across open areas

Avenues of Approach:
- Great Salisbury Road
- Reedy Fork Road

Key Terrain: Military crest of ridge west of Guilford Courthouse

Obstacles and Movement:
- Neither side employs man-made obstacles
- Sodden low ground at Horsepen and Hunting Creek inhibit fast movement
- Natural deadfalls in woods slow up cross-country movement

Cover and Concealment:
- Thick trees on second and third line offer both cover and concealment
- Shadows from late afternoon sun and gunpowder smoke reduce visibility
- Reverse slope on second line offers cover

</td></tr>
</table>

Figure 1–Battlefield effects Guilford Courthouse, 15 March 1781.
Harold Allen Skinner Jr.

the royal artillery. After Earl Cornwallis had consulted the guides concerning the nature of the country, and viewed as much as he could the disposition of the militia, he desired Major-general Leslie to move to the right with the 71st and the regiment of Bose, which force was to compose his front line for the attack on the enemy's left, and the 1st battalion of the guards was allotted for his reserve. Colonel Webster was directed to form the 23d and 33d on the left of General Leslie's division: Brigadier-general O'Hara instructed to support Colonel Webster, with the 2d battalion and grenadier company of the guards. While these troops were forming, the yagers and light infantry of the guards remained near the guns in the road; but when the line moved on, they attached themselves to the left of Webster's brigade. The artillery, under Lieutenant [Donald] Macleod, proceeded along the high road: The dragoons likewise could only move in column in the same direction . . . in reserve till the infantry could penetrate through the woods to the open ground. . . . During these arrangements for the attack, the British artillery cannonaded the enemy's center with considerable effect. . . . The troops were no sooner formed than they marched forward with steadiness and composure.[49]

Analysis

1. Before the battle, the Continental gunners used a local resident to estimate ranges to the most likely enemy avenue of approach. What other measures could the gunners have taken to place effective fire on the British corps?

2. Neither side used skirmishers during the opening stages of the battle. Was this a good or bad plan? Why?

3. Cornwallis seems to have ordered a straightforward frontal assault on the first American

---

[49] Tarleton, *A History of the Campaigns of 1780–1781 in the Southern Provinces of North America*, 272–73.

line, with no consideration of branch and sequel planning. As you go through the subsequent stands, pause to consider some mission-type orders or procedural controls Cornwallis could have used to better guide his subordinate commanders.

Stand 4: The American First Line Deployment

Directions

From the Hoskins House, return via vehicle to the parking lot of the Guilford Courthouse Battlefield Visitors Center. From the parking lot, move to the west of the visitor's center, then follow the sign for stop 1 on the NPS self-guided map, "The American First Line." After walking about 100 meters, you will arrive at "The Battle Begins" interpretive marker, which faces west, overlooking the paved New Garden Road.

Orientation

It is now around 1300 on 15 March 1781. Arrayed along a northwest-southeast running fence, facing west toward the British enemy at the Hoskins House, were two brigades of North Carolina Patriot militia, supported by a Continental artillery section with two 6-pounder guns sitting astride the road. Running to the north in order from left to right was Brigadier General John Butler's North Carolina militia brigade; Colonel Charles Lynch's Virginia rifle battalion; and Lieutenant Colonel William Washington's Corps of Observation anchoring the right flank. Running to the south in order was Colonel Thomas Eaton's North Carolina militia brigade; Colonel William Campbell's Virginia rifle battalion; and Lee's Partisan Corps anchoring the far left of the militia line. About 437 yards (400 meters) east of this position was a thick belt of timber that concealed the second line of defense, Brigadier General Edward Stevens's and Brigadier General Robert Lawson's Virginia State Line infantry brigades.

Description

Before entering the tactical discussion, the instructor should describe the effect of the terrain on ballistics and tactical movements. Troops armed with smooth-bore muskets tended to overshoot a target when firing uphill. Thus, Greene's positioning of the poorly trained North Carolina brigades facing westward on a slight downhill slope possibly was done to keep their shots low. Secondly, the muddy farm fields and intermediate fences to the west formed natural obstacles that would slow and disrupt the British movement into engagement range of the American line. Lastly, the contrast of the bright red British uniforms with the green vegetation farther to the west would help the American riflemen clearly identify and engage targets.

Facing the British along this line were Greene's short service militia, two brigades of around 1,000 North Carolina militiamen. A fair number of the North Carolina leaders and troops had seen active service in the various campaigns against Native American nations, but few had significant combat experience against British regulars. Compared to the all-volunteer Virginia Line infantry and Continentals, most North Carolina fighters were 90-day draftees, poorly trained and probably not enthusiastic to face combat. The troops were armed with a wide variety of firearms, including muskets, rifles, and shotguns. Few were issued bayonets, much less trained in their use, so the North Carolina brigades were poorly prepared for close combat against the British or German infantry.

Partially heeding Morgan's tactical recommendations, Greene anchored the flanks of the militia line with the corps of observation, and reinforced the line with an artillery section with two 6-pounders.

Significantly, Greene failed to follow several of Morgan's tactical innovations. First, no skirmishers were positioned forward of the main line to target officers and disrupt the enemy deployment for combat. Secondly, Greene deployed the North Carolina brigades well beyond mutual supporting distance of the Virginia Line brigades on the second line. Critically, Greene's orders to Brigadier General John Butler and Brigadier General Thomas Eaton were imprecise, only to "fire two volleys and then retreat," thus offering the North Carolina officers no guidance for further action.[50] Finally, after riding the ranks of the first line to give a last-minute exhortation to the North Carolina troops, Greene withdrew to the third line, leaving the militia brigadiers without an overall commander to control the first line fight.

Vignettes

Shortly before Brigadier General Daniel Morgan left the Southern Army due to debilitating sciatica and rheumatism, he offered his tactical thoughts on the coming battle to General Greene.

> DEAR SIR: I have been doctoring these several days, thinking to be able to take the field again. But I find I get worse. My pains are now accompanied by a fever every day. I expect Lord Cornwallis will push you until you are obliged to fight him, on which much will depend. You have, from what I see, a great number of militia. If they fight, you will beat Cornwallis; if not, he will beat you, and perhaps cut your regulars to pieces, which will be losing all our hopes. I am informed that among the militia will be found a number of old soldiers. I think it would be advisable to select them from among the militia, and put them in the ranks with the regulars; select riflemen also and fight them on the flanks, under enterprising officers who are aquainted with that kind of fighting; and put the militia in the center, with some picked troops to their rear, with orders to shoot down the first man that runs. If anything will succeed, a disposition of this kind will. I hope you will not look at this as dictating, but as my opinion on a matter that I am much concerned in.[51]

Just after Greene left the American first line for the rear, Lee's Corps of Observation rode in to the American lines after breaking off its running engagement with Tarleton's dragoons. Before riding to his designated position on the American left, Lee gave a brief exhortation to the North Carolina militia, as was recalled by an eyewitness:

> Cornwallis arrived at Guilford Court House between 12 and 1 o'clock on the 15th March 1781. Just before the battle commenced Col. Lee rode up to the lines where deponent stood and used something like these words, "My brave boys, your lands, your lives and your country depend on your conduct this day. I have given Tarleton hell this morning and I will give him more of it before night." And speaking of the roaring of the British cannon he said, "You hear damnation roaring over all these woods, but after all they are no more than we," and he went on to flank the left of the American army.[52]

Analysis

1. Was Cornwallis's force ratio compared to the Americans' sufficient for tactical and operational success? If not, what could Cornwallis have done to improve his odds for the attack?
2. How could Greene have used mission command-type orders and measures to guide the

---

[50] Babits and Howard, *Long, Obstinate, and Bloody*, 77.
[51] Conrad et al., *The Papers of General Nathanael Greene*, vol. 7, 324.
[52] Pension application of William Lesley, 1836, "Key Original Source Outline for the Battle of Guilford Courthouse."

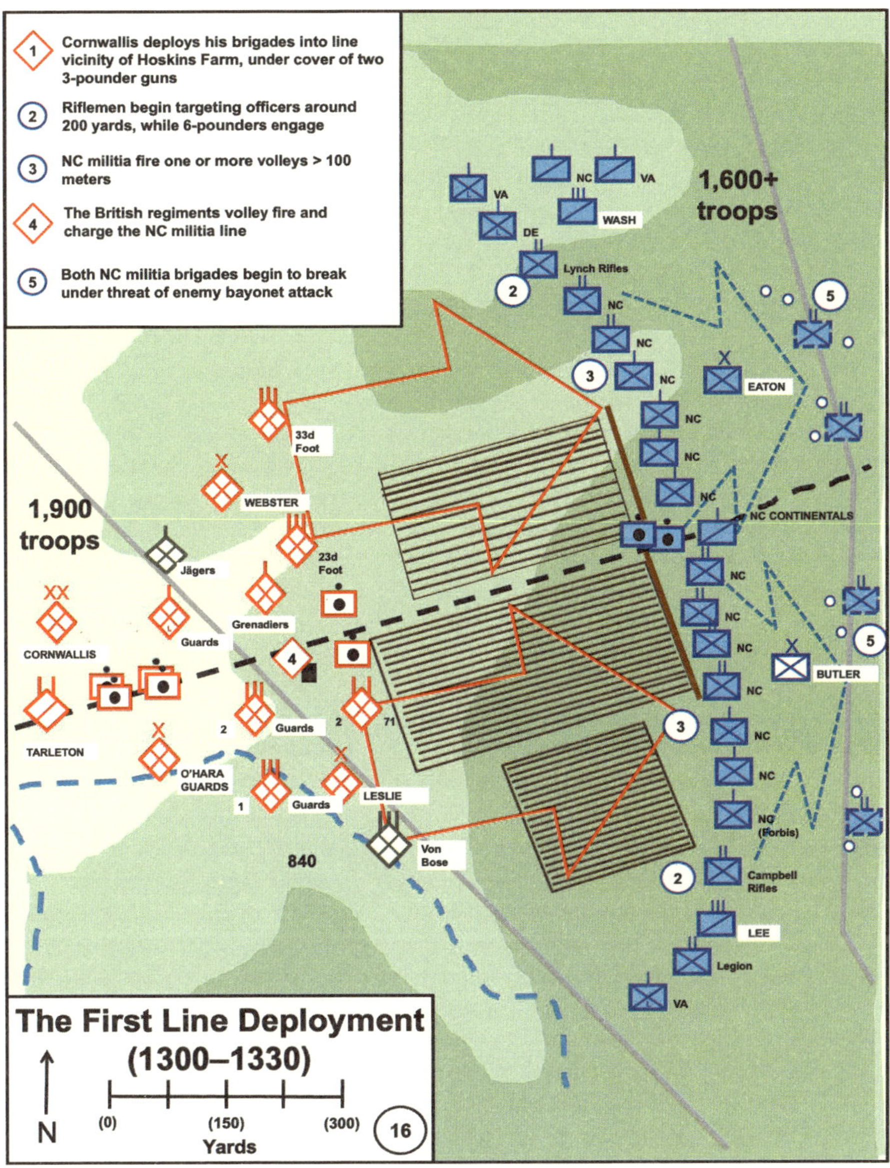

Map 16
Chuck Collins, Army University Press, and Harold Allen Skinner Jr.

first line battle in his absence? Think in terms of branch and sequel plans and decision points.

3. Compare Morgan's recommendations (and his tactical dispositions at the Cowpens) with Greene's actual plan. For what reasons might Greene have deviated from Morgan's advice? What risks did he assume? Also, what possible advantages might Greene have gained had he more closely followed Morgan's advice?

Stand 5: Breaking of the First Line

Directions

From stand 4, retrace your steps to the visitor's center to follow the curve of the sidewalk past the parking lot to the Battlefield Tour route (marked in red on the NPS trifold maps). Walk roughly 200 meters west then southwest to the NPS "Local Hero" marker and nearby (inaccurately marked) Colonel Arthur Forbis monument.

Orientation

This stand marks the approximate spot where Captain Arthur Forbis was mortally wounded when Eaton's North Carolina militia brigade was attacked by portions of Webster's and Leslie's brigades. Forbis's company was positioned on the extreme left flank of Eaton's brigade, which was in turn protected by Lee's Corps of Observation on the far left flank of the first line. Running to the north of Eaton's brigade was Butler's brigade, flanked by two companies of Continental light infantry and Washington's Corps of Observation screening the northern flank. Approximately 400 meters to the northeast of this location was the second American line, concealed in a thick belt of trees. About 75 meters due west of the Forbis marker is a white obelisk, which was erected after the war to commemorate Cornwallis's army.

Description

This stand covers the breaking of the southern half of the North Carolina militia line by the British, which took place at about 1315. Once Cornwallis ordered a general advance, Webster's *23d* and *33d Foot* marched forward toward this location with the *71st Foot* of Leslie's brigade in echelon to the right. During the movement to contact, hindered by muddy fields and at least one fence, British company noncommissioned officers shouted and pushed their troops, keeping the ranks in alignment to prevent American units from enfilading or flanking the line. North Carolina riflemen in the ranks and the Virginia riflemen on the right flank engaged enemy officers once the British line approached within 300 meters or so. The sporadic rifle fire did not slow the seemingly inexorable advance of the British line, and many of the green Carolina militia, already unnerved by the British cannon fire, began to fire despite the orders of their officers. The irregular fire from the North Carolina line caused enough casualties for Webster and Leslie to hurry the pace of their troops across the danger zone.

Webster's brigade spontaneously stopped for a moment when they realized the American first line was much larger than expected, as Roger Lamb recounted: "A general pause took place; both parties surveyed each other a moment with the most anxious suspense," before Webster rode to the front of the *23d Foot*'s line and ordered his brigade to resume the movement at the quick-step.[53] The British line quickly advanced to close proximity of the American line, absorbing at least one or more coordinated volleys from the North Carolina line without pausing to return fire. At the killing distance of 40 meters, Webster halted his brigade and ordered a single volley before a bayonet charge—a tactic designed to shock the Americans sufficiently to prevent a point-blank return volley before the British could close into bayonet range. Despite the British volley, a few American militia companies managed to reload and fire a second volley at near point-blank range. For the most part, Webster's assault succeeded as intended, as Eaton's North Carolina brigade broke after a brief skirmish with bayonets and clubbed muskets, where the militia "got panick struck and ran from the scene of action."[54] Lee's Corps of Observation on Eaton's left flank remained in position to enfilade Webster's left flank as it passed through the first line position to continue the attack. The fragmenting of the American line and resulting enfilade fire from

<hr>

[53] Lamb, *An Original and Authentic Journal of Occurrences during the Late American War*, 361.
[54] Babits and Howard, *Long, Obstinate, and Bloody*, 112.

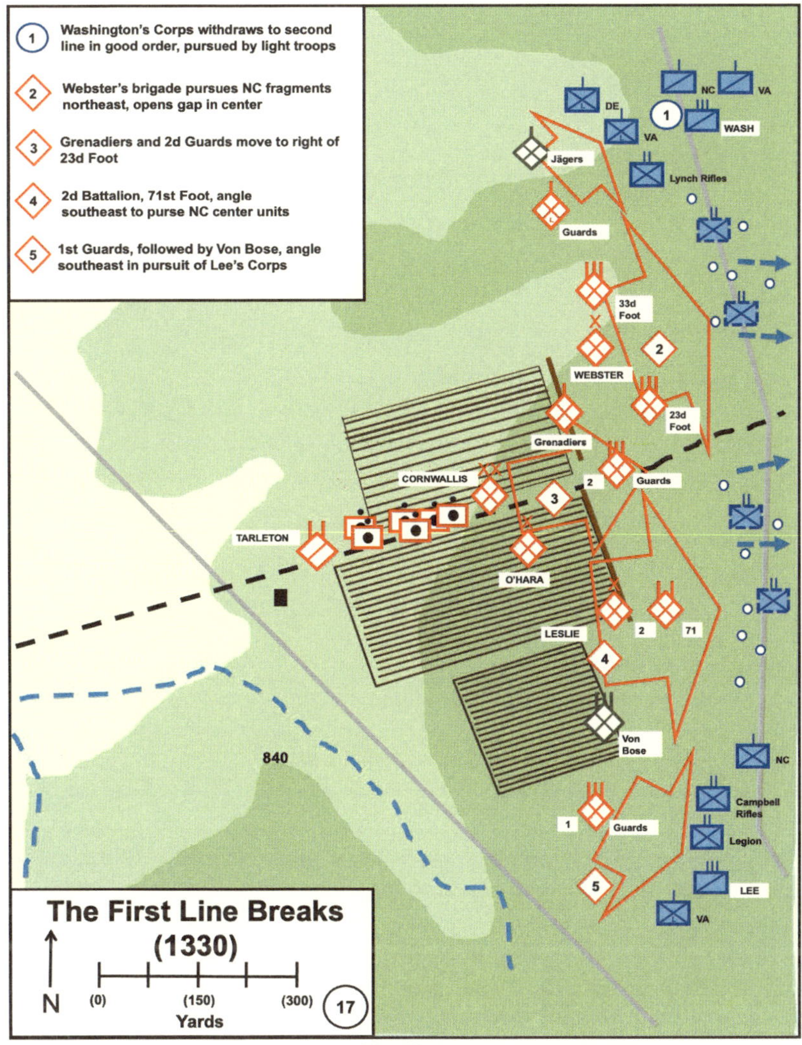

Map 17

Chuck Collins, Army University Press, and Harold Allen Skinner Jr.

Lee's corps caused Webster's and O'Hara's brigades to angle northeast, while Leslie's brigade on the British right reoriented toward Lee's Corps of Observation on the left, a move that opened a sizable gap between the British left and right wings.

Although this was Forbis's first action, Forbis seemed to have been a well-respected commander who led his troops by example. Faced with the forbidding wall of British bayonets, Forbis demonstrated his resolve by killing a British officer with a well-aimed rifle shot. Forbis's company exercised good fire discipline in engaging the British line, and their first volley caused many casualties in the *71st Foot*. When faced with the British bayonet assault, Captain Forbis kept his company in position and continued to fight until only four unwounded soldiers were left to flee the battle. Forbis was mortally wounded at this spot, knocked to the ground by two musket balls, and bayonetted as he lay wounded on the ground.

Vignettes

One of the North Carolina militia commanders who survived the battle was Colonel James Martin, commander of the Guilford County militia regiment:

> I was posted in the front line with scarce a complete Captain's company commanded by Captain Forbis, a brave undaunted fellow. We were posted behind a fence and I told the men to sit down until the British who were advancing came near enough to shoot. When they came in about 200 yards I saw a British officer with a drawn sword driving up his men. I asked Capt. Forbis if he could take him down. He said he could for he had a good rifle and asked me if he should shoot then. I told him to let him in 50 yards and then take him down, which he did.[55]

Private James Collins, a teenaged veteran of the Kings Mountain battle and a rifleman in the Franklin County, North Carolina, militia regiment, remembered that

> [He joined] his company while drawn up in line of battle, soon after which the battle of Guilford Court House commenced. . . . He was placed in the first line . . . about 200 yards to the right of an open field, and when the British made their charge he saw the disgraceful retreat of that portion of the militia which was placed behind the fence of the field. He, with most of his company, stood till they gave four fires, when finding the retreat pretty general, he also fell back and retreated to the Iron Works.[56]

Commissary General Stedman described the first line fight:

> As soon as the head of the British column appeared in sight of the first line of the Americans, a cannonade was begun from the two six-pounders, posted upon the center of the road, which was immediately answered by the royal artillery under lieutenant Macleod; and whilst this cannonade continued, the British commander, with the utmost dispatch, made his dispositions for the attack . . . [detailed list of the British order of battle] . . . this disposition being made, the line received orders to advance and moved forward with that steady and guarded, but firm and determined resolution which discipline alone can confer. It has been remarked by an eye-witness that "the order and coolness of that part of Webster's brigade which advanced across the open field, exposed to the enemy's fire, could not be sufficiently extolled." At the distance of one hundred and forty yards they received the enemy's first fire, but continued to advance unmoved. When arrived at a nearer and more convenient distance, they delivered their own fire, and rapidly charged with their bayonets: The enemy did not wait the shock, but retreated behind the line. In other parts of the line the British troops behaved with equal gallantry, and were not less successful.[57]

Analysis

1. How did the perception of the American militia affect how Greene and Cornwallis directed the battle?
2. Did the militia meet their specified and implied tasks as given by General Greene?
3. How could Greene have better influenced the outcome of the first line fight?
4. How could the North Carolina militia officers have improved the effectiveness of their fires? What measures could they have taken to reduce their vulnerability to enemy bayonets?

---

[55] James Martin, pension application with written account of the Guilford Courthouse battle, Documenting the American South (website), Colonial and State Records of North Carolina, vol. 22, 17 October 1832, 148.

[56] C. Leon Harris, transc., "Southern Campaign American Revolution Pension Statements: Pension Application of James P. Collins W6737" (PDF), 12 September 1832, transcribed 31 October 2014, Southern Campaigns Revolutionary War Pension Statements and Rosters (website), accessed 30 March 2020.

[57] Stedman, *The History of the Origin, Progress, and Termination of the American War*, vol. 2, 338–39.

Stand 6: The Fragmented Attack
Directions

From the Forbis monument, follow the paved foot trail that parallels the paved vehicle route (red route) about 300 meters around the curve to stop 2 on the NPS tour map. Once there, orient students to the NPS interpretive panel adjacent to a small parking lot north of the vehicle route titled the "Fragmented Attack."

Orientation

Stand 6 is at the approximate location for the left flank of Brigadier General John Butler's North Carolina brigade. From the southwest was the general direction advance for Major General Leslie's brigade. The center of the North Carolina line was to the northwest, while the left flank of the American second line was to the east. To the south and east of this location was Lieutenant Colonel Lee's Corps of Observation, securing the far left of the American line.

Description

This stand discusses the opening stages of the running battle between the British right flank units and Lee's corps and the cohesive fragments of the North Carolina militia line—a fight that began around 1315. Fire discipline in Butler's North Carolina militia brigade was significantly better than Eaton's brigade, as evidenced by the high number of British and Hessian casualties resulting from the engagement. Contributing to the toll was Leslie's decision to alternate fire and movement, a tactic that prolonged the exposure of his troops in the American engagement zone. However, once the *71st Foot* and *Von Bose Regiment* had closed to bayonet range, many militiamen had little choice but to scatter and run for the rear. Yet, several North Carolina companies on the far left maintained their cohesion and fell back to join Colonel William Campbell's Virginia rifle battalion. As Leslie's brigade pushed through the fleeing North Carolinians, Lee's Corps of Observation began firing and maneuvering, trading space for time in a bid to avoid a decisive engagement with the enemy infantry. Lee's maneuvering to the southeast drew the *Von Bose Regiment* in pursuit, a movement that opened a large gap between the German troops and the supporting *71st Foot.* Either Cornwallis or Leslie (sources are uncertain) spotted the gap and ordered the *1st Guards* battalion forward to cover the vulnerable left flank of the *Von Bose Regiment.* The *Von Bose* and *1st Guards* battalion continued to arc southeast away from the principal axis of advance, fighting sharp engagements with platoons of North Carolina militia while seeking to come to grips with Lee's Corps of Observation.[58]

Vignettes

Captain Thomas Saumarez commanded a wing of the *23d Foot* at Guilford Courthouse. He described the difficulties faced during the movement to contact with Butler's North Carolina brigade:

> The Royal Welch Fusiliers [*23d Foot*] had to attack the enemy in front, under every disadvantage, having to march over a field lately ploughed, which was wet and muddy from the rains which had recently fallen. The regiment marched to the attack under a most galling and destructive fire, which it could only return by an occasional volley.[59]

Captain Henry Connelly was one of General Butler's company commanders. Having previously led militia at other battles, including the Cowpens, Connelly was in turn chagrined and deeply angered by the conduct of his troops.

---

[58] The road used by Lee's dragoons is no longer visible but probably ran parallel to the modern Old Battleground Road.

[59] Stouppe McCance and A. D. L. Cary, *Regimental Records of the Royal Welch Fusiliers, 1689–1918,* vol. 1 (London: Foster Groom, 1921), 180.

After this engagement [the Cowpens] we all formed a junction with Genl. Greene [and] marched back to Guilford Court house, and this applicant actively participated in that memorable battle, and he had the great mortification to see his men in a panic fly at the approach of the enemy; and although this applicant endeavored to rally them, it was impossible, and many even retreated to their homes, but this applicant remained and continued to fight until the Americans were thrown into disorder and confusion & defeated.[60]

Private Nathan Slade of Caswell County, North Carolina, experienced his first battle at Guilford Courthouse and lived to tell his story.

At the battle of Guilford, I was one of the North Carolina militia. We were in that battle stationed by Genl. Greene behind a fence that being a position which he thought most advantageous for raw troops who were unaccustomed to stand the shock of battle. We were ordered to stand firm and to do what we could do. The Genl. who had the immediate command of that portion of the troops to which I belonged was [Genl.] Butler—an officer in whose courage and skill I had then no confidence and have now but little if any respect for his character. The enemy approached us and were according to the best of my belief within eighty to an hundred yards of us when they made their first fire—my recollection is that most of us stood firm until after the second fire. On the third fire there were but few if any of us left to receive it—all or nearly all had broke and retreated in great disorder.[61]

Analysis

1. By all accounts, the officers and soldiers of Eaton's North Carolina brigade performed better than those of Butler's brigade. What are some possible reasons for the difference?
2. Withdrawing while under pressure is considered a difficult and dangerous tactical task. What are some ways the American commanders could have prepared the troops to perform such a task?
3. What are some reasons that Lee might have withdrawn his corps away from the rest of the American army?

Stand 7: Lawson's Brigade of the Virginia State Line
Direction

From the "Fragmented Attack" NPS marker, move east to follow the paved road (red route) until it intersects the foot trail that runs parallel to Old Battleground Road. Follow the foot trail north, past the visitor's center and to the graveled New Garden Road path. Turn right (east), cross Old Battleground Road, and continue to walk east to the vicinity of the Greene monument. Here you find an intersecting paved trail; follow the cross trail to the north about 25 meters to a spot under some large shade trees, a spot that marks the approximate center point of the second American line.

Orientation

This stand is located at the approximate left flank position of General Robert Lawson's Virginia Line brigade, which was positioned in a line running to the northeast. The line of Brigadier General Ed-

[60] Will Graves, transc., "Southern Campaign American Revolution Pension Statements: Pension Application of Henry Connelly W8188" (PDF), 15 August 1833, transcribed 1 December 2010, Southern Campaigns Revolutionary War Pension Statements and Rosters (website), accessed 20 June 2017.
[61] Will Graves, transc., "Southern Campaign American Revolution Pension Statements: Pension Application of Nathan (Nathaniel) Slade W6071" (PDF), June 1832, transcribed 14 May 2009, Southern Campaigns Revolutionary War Pension Statements and Rosters (website), accessed 7 June 2017.

ward Stevens's Virginia Line brigade began on the other side of the Salisbury Road and continued southeast into what is now the Forest Lawn Cemetery. To the northeast about 650 meters was the location of the main Continental line. The avenue of approach used by Lieutenant Colonel Webster's brigade was to the southwest. Just south of this location is a large monument commemorating General Greene's leadership of the Southern Department during the American Revolution.

Description

Although Lawson commanded only three understrength regiments, around 650 men, the Virginia Line troops had a greater degree of combat readiness as compared to the conscripted North Carolina militia. A large proportion of the Virginia officers and troops had previous combat experience during the revolution, and as a whole the regiments were better trained and more cohesive due to their longer periods of enlistment. Brigadier General Lawson was an experienced combat leader, having commanded the 4th Virginia Continental Regiment during the Trenton and Princeton campaigns in 1776–77.

Lawson's three regiments were deployed as follows: Colonel John Randolph's unit held the American right flank; Colonel John Holcombe's regiment held the middle, and Major Henry Skipworth's understrength regiment secured the brigade left. As in the first line, the officers did not order the soldiers to fortify their positions, but the soldiers were allowed to disperse in open order to use trees as cover. Lawson deployed a line of skirmishers in the woods to provide early warning, an unnecessary precaution as the noise of battle and rush of fleeing Carolinian militia gave ample notice of the British advance. In reacting to Webster's approach, Lawson overestimated the abilities of his troops when he ordered Randolph and Holcombe to attack the seemingly vulnerable *23d Foot*. The northward swing of the two Virginia regiments left Holcombe's left flank uncovered, a vulnerability exploited by a well-timed attack by the *Guards Grenadier Company*, which was in echelon to the right of the *23d Foot*. Near the road, Skipworth's tiny regiment was hit in front by the *23d Foot* and on the right flank by the *Guards Grenadier Company*. The resulting collapse of Holcombe's regiment led to the rapid disorganization of Lawson's entire brigade. However, many of Lawson's small unit leaders managed to reform squads and platoons of Virginia soldiers into a jumbled skirmish line. Despite the disparity in numbers, the Virginia troops fought fiercely, killing many redcoats and seriously wounding Brigadier General O'Hara. At one point, the Virginia resistance was so strong that Cornwallis's horse was killed by a musket ball as he personally led some British troops into a skirmish. Ultimately, Lawson's brigade failed to stop the forward advance of the more cohesive British battalions, and the surviving Virginia platoons fought a delaying action to the northeast to fall in with Washington's Corps of Observation. Casualties in the close-range fighting were heavy on both sides, and American officer losses were particularly heavy due to the deadly accurate fire of the Hessian jägers.

Vignettes

Virginia militia Major St. George Tucker described the collapse of Lawson's attack.

> A cannonade of half an hour ushered in the battle. Our friend Skipwith was posted in the express direction of the shot, and, with his battalion, maintained his post during a most tremendous fire with a firmness that does him much honor. Col. Holcombe's regiment was on the right of him and on my left, so that I was in perfect security during the whole time, except for a few shot which came in my direction. . . . When the cannonade ceased, orders were given for Holcombe's regiment, and [Randolph's] on the right . . . to advance and annoy the enemy left flank. While we were advancing to execute this order, the British had advanced, and, having turned the flank of [Skipwith's regiment], we discovered

them [the *23d Foot*] in our rear. This threw the militia into such confusion . . . Holcombe's regiment and ours instantly broke off without firing a single gun, and dispersed like a flock of sheep frightened by dogs. With infinite labor, [Major John] Beverley and myself rallied about sixty or seventy of our men . . . we at several times sustained an irregular kind of skirmishing with the British, and were once successful enough to drive a party for a very small distance.[62]

Virginia private John Chumbley survived to write of his first experience of battle.

In February 1781 he was again drafted in the said county of Amelia and was attached to a company commanded by Capt. Robert Hutson [Robert Hudson] . . . the company was supplied with arms, ammunition, and provisions. Thence he marched in said Company 20 miles . . . where they joined four other companies [in] Lawson's Brigade. . . . In the front of the Army . . . was placed the North Carolina militia. About 200 or 300 yards, in the rear of these were stationed the Virginia militia in a line at right angles with the road on the right. . . . The first [line] gave way and retreated on the appearance of the enemy, who, after a severe fire from our [flanking] parties in advance of the second line, advanced directly upon the Virginia Militia which were mostly placed under the covert of an under growth of [word obscured]. This line withstood the enemy for some time under a severe fire, and the Virginia Militia certainly did great execution in this encounter, but were finally routed completely; and the right wing . . . gave way first, and perhaps did not act so gallantly, nor fight so obstinately as the left.[63]

Sergeant Roger Lamb, a noncommissioned officer in the *23d Foot*, provides a British viewpoint of the fight with the Virginia Line regiments.

At last the Americans gave way, and the brigade advanced, to the attack of their second line. Here the conflict became still more fierce. Before it was completely routed . . . I observed an American officer attempting to fly. I immediately darted after him. . . . when, hearing a confused noise on my left, I observed several bodies of Americans drawn up within the distance of a few yards. . . . I stopped and replenished my own pouch with [cartridges] . . . several shots were fired at me; but not one took effect. Glancing my eye the other way, I saw a company of the guards advancing to attack these parties. . . . It was impossible to join this company, as several of the American parties lay between me and it. . . . I saw lord Cornwallis riding across clear ground. . . . mounted on a dragoon's horse (his own having been shot;) the saddle-bags were under the creature's belly, which much retarded his progress . . . his lordship was evidently unconscious of his danger. I immediately laid hold of the bridle of his horse. . . . then mentioned to him, that if his lordship had pursued the same direction, he would in a few moments have been surrounded by the enemy. . . . I continued to run along side of the horse . . . until his lordship, gained the 23d regiment, which was at that time drawn up in the skirt of the woods.[64]

[62] "The Southern Campaign, 1781, from Guilford Court House to the Siege of York, Narrated in the Letters from Judge St. George Tucker to His Wife, the Southern Campaign as Narrated by St. George Tucker," 40.

[63] Will Graves, transc., "Southern Campaigns American Revolution Pension Statements: Pension Application of John Chumbley S32169" (PDF), 15 May 1833, transcribed 6 November 2014, Southern Campaigns Revolutionary War Pension Statements and Rosters (website), accessed 7 June 2017.

[64] Roger Lamb, *An Original and Authentic Journal of Occurrences during the Late American War, from Its Commencement to the Year 1783* (Dublin, Ireland: Wilkinson and Courtney, 1809), 361–62.

Analysis

1. No extant account exists of General Greene's exact orders to Brigadier General Lawson for the second line defense. Using *Offense and Defense*, paragraphs 4-42 through 4-47 "Mission Command," or *Commander's Tactical Handbook*, section on "General Coordinating Instructions for Defense Orders," consider some possible mission command-type orders Greene could have issued for the second line defense.[65]
2. What are some ways Lawson could have better prepared his regiments for battle? What of General O'Hara?
3. What are some ways Greene could have minimized risk to the Virginia militia?
4. Contrast Cornwallis's and Greene's locations and involvement during this phase of the battle; consider the advantages or disadvantages.

Stand 8: Stevens's Brigade of the Virginia State Line
Directions

Turn south and retrace your steps to the Salisbury Road trail. From that intersection, turn left (east) and walk about 30 paces east on the gravel trail to a brick-paved path that passes in front of the massive General Greene monument to the right. Turn right (south) and follow the brick path, which turns into an asphalt-paved trail. Continue to follow the asphalt foot trail south and east to a Y intersection. Follow the right fork southeast and follow the gravel path to a brown NPS sign for stop 3 on the NPS tour map—"Sustained Firefight." This stop is just to the east of a small parking lot on the vehicle/bicycle tour road.

Orientation

The location for stand 8 marks the approximate middle of Brigadier General Stevens's brigade. After arriving at this point, first face the participants to the NPS interpretive panel titled "Sustained Firefight." To the east-northeast approximately 750 meters was the position of the Continental third line. To the west, the general axis of advance for the approaching regiments of Leslie's brigade, while to the north and west was the line of Lawson's Virginia brigade. The line of Stevens's brigade continued to run southeast for another 200 meters, which today is within the grounds of the Forest Lawn Cemetery. Out of respect for legal and cultural concerns, the cemetery is not included in this guide.

Description

With the splintering of the British main battle line, the fight at the second line evolved into two distinct engagements involving Lawson's and Stevens's brigades. This stand reviews the other major part of the second line fight starting around 1345, Stevens's Virginia State Line brigade against elements of Leslie and O'Hara's brigades. During the walk to this stand, point out how the terrain slopes eastward from the American second line. Accounts of the battle suggest the Virginia troops were deployed in a reverse slope defense along the visible ridgeline. As in Lawson's brigade, the regiments of Stevens's brigade had greater combat efficiency than the North Carolina militia. Stevens had a significant amount of Continental combat experience, having served as the commander of the 10th Virginia Continental Regiment during the Brandywine and Germantown campaigns, before resigning his commission in 1778. In addition, Stevens wanted revenge for his humiliating defeat at Camden, where his Virginia State Line brigade had fled the battlefield in panic, leaving Johann de Kalb's Continentals open to destruction by Cornwallis's army.

---

[65] *Offense and Defense*, 4-7–4-8; and *Commander's Tactical Handbook*, 51.

Stevens arrayed his Virginia State Line regiments (from left to right: Major Alexander Stuart's Rockbridge and Augusta Counties, Colonel George Moffett's Augusta County, Colonel Nathaniel Cocke's Lunenburg and Halifax Counties, and Colonel Peter Perkins's Pennsylvania County) in open order to take full advantage of a reverse slope defense. Stevens's flexible deployment left room for maneuvering and space for the expected withdrawal of the North Carolina line. Before the battle, Stevens exhorted his troops to stand firm and defeat the British, and he backed up his threat to execute deserters by posting a line of riflemen with express orders to shoot deserters.

Facing Stevens's brigade were elements of the *2d Guards*, the *71st Foot*, the *1st Guards*, and on Stevens's extreme left, the *Von Bose Regiment*. The cohesion of the regiments had been affected by the first line, and the effects were magnified during the movement to contact by the thick vegetation and gunpowder smoke. As a result, the fight on the second line evolved into a swirl of confusing small unit actions, marked with bloody close-range volleys followed by brutal hand-to-hand fighting with bayonets, clubbed muskets and rifles, and edged weapons. After his horse was killed, Stevens continued to ably lead his brigade on foot for several minutes until a British ball shattered his femur. In the resulting collapse of morale, only Major Alexander's regiment remained cohesive enough to withdraw to link up with Lee's Corps of Observation as it fought a delaying action to the southeast against the *1st Guards* and *Von Bose*. To the far right, several Virginia platoons remained intact and fought to the east to fall in with the Continental third line. Stevens's brigade had been shattered, but not before inflicting heavy casualties on Cornwallis's army.

Vignettes

Lieutenant Colonel Lee's hyperbolic description of Stevens's second line fight:

> The front line of Greene, although advantageously posted and gallantly flanked . . . were struck with a panic at the first fire . . . threw down their loaded arms, and fled precipitately. . . . But the Virginians, not dismayed, but indignant, stood firm against the wave of flight and battle that rolled impetuously toward them; and the action here became close and animated. Nothing could surpass the ardour of the British; the resistance of the Virginians was rude and resolute. . . . [After the collapse of Lawson's brigade] the centre however, still maintained itself; and the spirit of Stevens seemed to expand, as the battle swelled, like the oak which displays its strength and its volume in proportion as the winds arise. More determined in danger, he saw without shrinking, fragments of the war fall off from his right . . . he maintained the fight with vigour and vivacity. At length, in a furious charge by the enemy, a ball pierced his thigh, and this gallant officer, no longer able to stand foremost in danger, ordered a retreat; leaving his adversaries too much weakened by the conflict to exult in their success.[66]

British Commissary General Stedman recounted his perspective of the second line fight.

> The enemy [first line] did not wait the shock, but retreated behind their second line. . . . The second line of the enemy made a braver and stouter resistance than the first. Posted in the woods, and covering themselves with trees, they kept up for a considerable time a galling fire, which did great execution. At length, however, they were compelled to retreat, and fall back upon the continentals. In this severe conflict, the whole of the British infantry were engaged: General Leslie, from the great extent of the enemy's front, reaching far beyond his right, had been very early obliged to bring forward the first battalion of the guards . . . and

---

[66] Henry Lee, *The Campaign of 1781 in the Carolinas; with Remarks Historical and Critical on Johnson's Life of Greene* (Philadelphia, PA: E. Littell, 1824), 171–73.

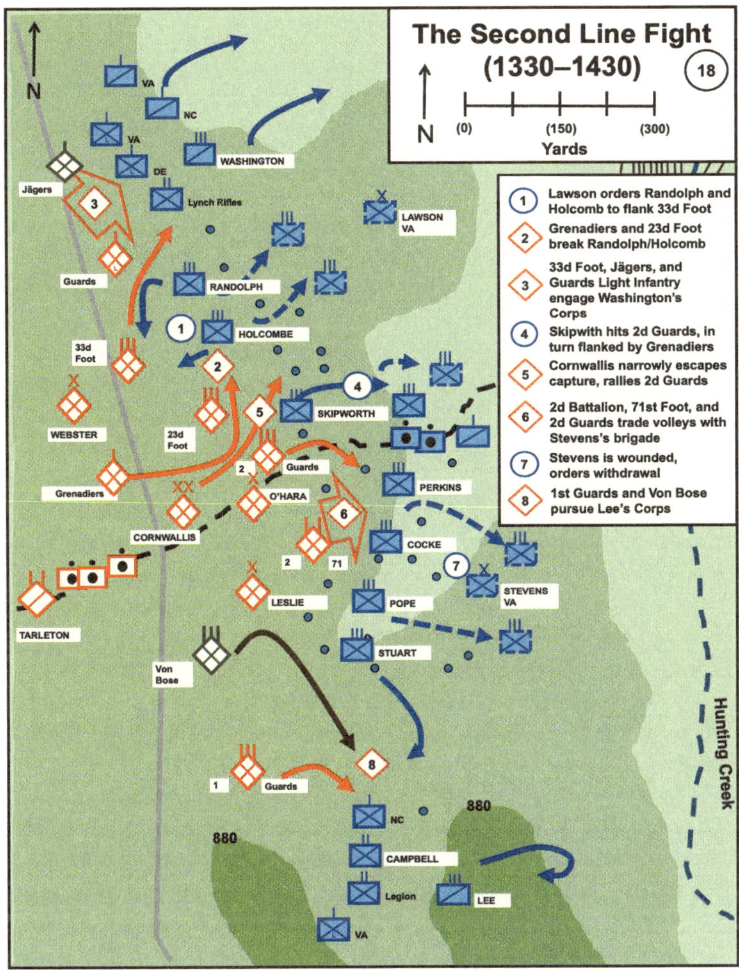

Map 18
Chuck Collins, Army University Press, and Harold Allen Skinner Jr.

form it into line. . . . The British line, being so much extended . . . was unavoidably broken into intervals in the pursuit of the first and second American lines . . . in consequence of the different degrees of resistance that had been met with, or of other impediments arising from the thickness of the woods, and the inequality of the ground.[67]

A soldier's perspective of the second line fight was left by Reverend Samuel Houston, who seemed to have fought in the ranks rather than fulfilling a chaplain's role.

Was rainy in the morning. . . . About ten o'clock, lying about our fires, we heard our light infantry and cavalry . . . begin firing with the enemy. Then we immediately fell into our ranks, and our brigades marched out. . . . Col. [Joseph] McDowell's battalion of Gen. Stephens' brigade was ordered on the left wing. When we marched near the ground we charged our guns. Presently our brigade major came, ordering to take trees as we pleased. The men run to choose their trees, but with difficulty, many crowding to one, and some far

[67] Stedman, *The History of the Origin, Progress, and Termination of the American War*, vol. 2, 339–40.

behind others. . . . Standing in readiness, we heard the pickets fire; shortly the English fired a cannon, which was answered; and so on alternately, till the small armed troops came nigh; and then close firing be[g]an near the centre. . . . and spread along the line. Our brigade major, Mr. [John] Williams, fled. . . . Soon the enemy appeared to us; we fired on their flank, and that brought down many of them; at which time Capt. [Alexander] Tedford was killed. We pursued them about forty poles, to the top of a hill where they stood. . . . And they a second time made us retreat back to our first ground.[68]

Analysis

1. Consider Brigadier General Stevens's implied mission in terms of *Offense and Defense* or *Commander's Tactical Handbook*, "Defensive Operations" chapter: Was his key tactical task *area defense* or *retrograde*? How could defining the mission differently influence the course and the outcome of the second line fight in Steven's sector?[69]
2. Reverend Houston's account graphically describes a confused small unit action in limited visibility. Given the limited time the Virginia brigades had before the battle, what could the Virginia platoon leaders and company commanders have done to better prepare their units for battle?

Stand 9: The Fight in the South

Directions

From the "Sustained Firefight" marker, move south, then walk east on the paved vehicle tour road about 300 meters until the road bends to the northeast. During the walk to the east, pause at the clearly visible creek to the north, an excellent vantage point to discuss the Virginians' use of a reverse slope defensive. Also, point out the large number of dead trees that serve as a good example of natural obstacles like those described by Commissary Stedman's vignette back at stand 8. At the bend in the road, cross to the right (south) to a small parking area and the "Expanding Battle" interpretive panel, which is marked as stop 4 on the NPS tour map.

Orientation

Stand 9 is at the approximate center point of the separate action in the southeast, which lasted roughly from 1430 to 1600. Once oriented south to the NPS marker, point to the west, which indicates the general direction of advance for Leslie's brigade in pursuit of Lee's Corps of Observation. From this location, the final engagement between the Virginia riflemen and Tarleton's dragoons ended at a location about 600 meters due south of this spot, in an area now part of the Greensboro County Park. To the north-northeast approximately 500 meters was the Continental third line. Here, it is useful to point out the open terrain, which was suitable for mounted operations by Tarleton and Lee's dragoons. The instructor also should note that the road used by Lee's troops in this area no longer exists, likely destroyed in the making of Forest Lawn Cemetery. Adjacent to this stand are postwar internment markers that are not included in this stand.

---

[68] Will Graves, transc., Thursday, March 15th entry, "Southern Campaign American Revolution Pension Statements and Rosters: Journal of Samuel Houston" (PDF), Southern Campaigns American Revolution Pension Statements and Rosters (website), 11 December 2011, accessed 15 June 2017. The unit of measure *poles* refers to a unit of length used by a surveyor, a *rod*, which could range between 3 and 8 yards. In modern usage, a rod is 16.5 U.S. survey feet; 40 poles is approximately 660 feet.

[69] "Area defense is a defensive task that concentrates on denying enemy forces access to a designated terrain for a specific time, rather than destroying the enemy outright," while a retrograde is a "defensive task that involves organized movement away from the enemy." The three forms of retrograde are: delay, where a force under pressure trades space for time, damaging the enemy while avoiding a decisive engagement; withdrawal, a planned retrograde where the force in contact disengages and withdraws out of combat range; and retirement, which applies to an otherwise unengaged unit. *Offense and Defense*, 4-3–4-4; and *Commander's Tactical Handbook*, 73–82.

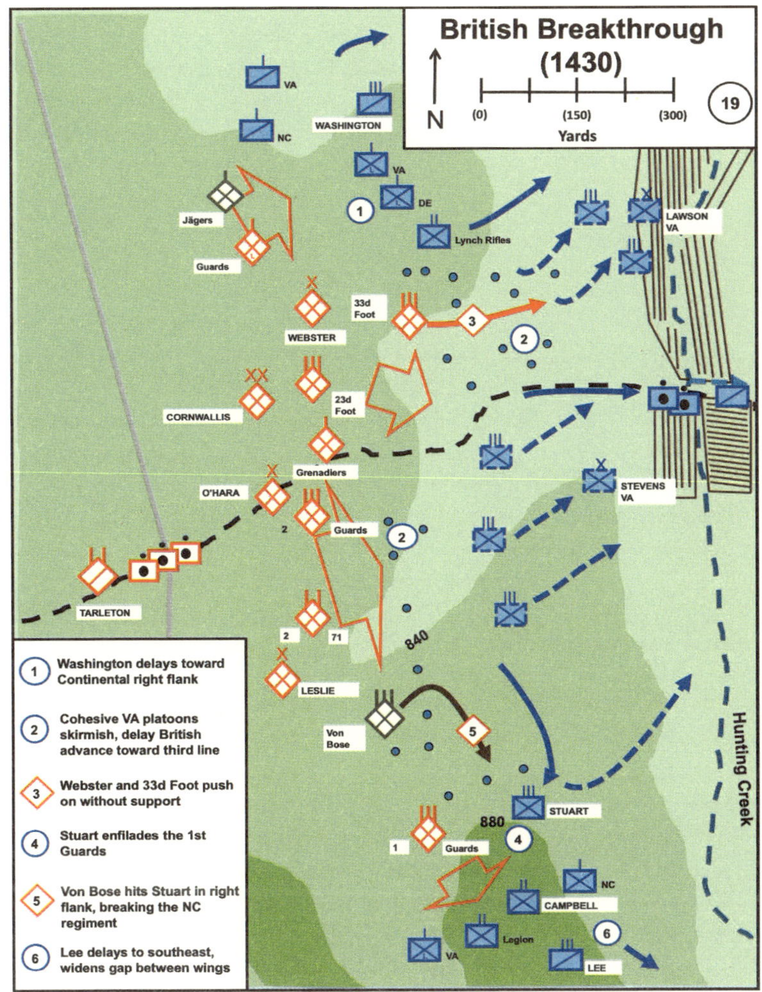

Map 19
Chuck Collins, Army University Press, and Harold Allen Skinner Jr.

Description

This stand will review the "battle within the battle," between Lee's reinforced Corps of Observation, the *1st Guards* battalion, and *Von Bose Regiment*. Part of the reason for the development of the separate fight was Lee's alignment on a small road that ran southeast of the Salisbury Road. The use of the road allowed for easy maneuvers by Lee's troopers but also served to draw a sizable portion of both armies into a peripheral battle that did little to affect the outcome of the main fight.

As the *1st Guards* battalion advanced in open order after Lee's corps, Lee's riflemen fought an effective delaying action by using the dense woods to maintain a stand-off distance from the British bayonets. Many British officers were killed or wounded by shots from the sides and rear, and the *Guards* were so battered that "it was at last entirely broken."[70] Fortuitously, the blue-coated Hessians entered the engagement, surprising the Americans and giving the *Guards* leaders time to rally their companies.

---

[70] Stedman, *The History of the Origin, Progress, and Termination of the American War*, vol. 2, 341.

Major Johann du Buy, the Hessian regimental commander, cannily avoided the vulnerabilities of the British open formation by keeping his German troops in compact columns. As the *Von Bose Regiment* pushed on toward the *1st Guards* battalion, the American skirmishers maneuvered to strike the enemy flanks and rear. Major du Buy responded by maneuvering his infantry companies in back-to-back lines, thus forming a mobile hedgehog perimeter with bayonets keeping the Americans at bay. Firing was intense, and both sides suffered heavy casualties in the close-range firefight. Compounding the horror of the fight, smoldering cartridge papers triggered a fire in the dry underbrush, and several of the wounded were roasted to death in the flames.

The fight on the flank actually outlasted that of the third line (covered in subsequent stands) and was brought to a close by two circumstantial events. First, Lee unexpectedly broke contact with his Legion dragoons and infantry toward Guilford Courthouse. Lee's surprise withdrawal left the attached Virginia and North Carolina light infantry vulnerable when a squadron of Tarleton's dragoons arrived on the scene. Several of the Americans were sabered down as they attempted a fighting withdrawal, and one American scored a lucky hit, maiming Tarleton's right hand with a rifle ball. By arcing to the northeast, Lee successfully avoided further enemy contact and rejoined the Continental force near Speedwell Furnace.

Vignettes

First, Lee's account of his mission:

> On the right, Lieutenant-Colonel Washington, with his cavalry . . . was posted, with orders to hold safe that flank. For the same purpose, and with the same orders, Lieutenant-Colonel Lee was stationed on the left flank with his Legion and the Virginia riflemen commanded by Colonel Campbell.[71]

Next, Lee's account of his withdrawal:

> When Lieutenant Colonel [Chapple] Norton [commander of the *1st Guards*] determined to unite with that part of the British line, which by successive detachments had reached and engaged the continentals. He therefore drew off in that direction, and all apprehensions of a defeat in this quarter being removed by his disappearance, Lieutenant Colonel Lee directed his cavalry to repair to the left of the Continentals, there to act until further orders; and turning with his infantry upon the regiment of Boze, with which the riflemen were engaged, the Germans fell back, and were pursued by Colonel Campbell; when Lee with his infantry, and one company of riflemen, pressed forward to join the continentals, and take his appropriate station on their left. In his progress he again encountered the guards under Norton, and passing to the right of the British, after Greene had retreated, joined his cavalry near the court-house.[72]

Contrast Lee's account with that of Commissary General Stedman:

> At one period of the action the first battalion of the guards was completely broken. It had suffered greatly in ascending a woody height to attack the second line of the Americans, strongly posted upon the top of it, who, availing themselves of the advantages of their situation, retired, as soon as they had discharged their pieces, behind the brow of the hill, which protected them from the shot of the guards, and returned, as soon as they had loaded. . . . Notwithstanding the disadvantage . . . the guards reached the summit of the

---

[71] Henry Lee, *Memoirs of the War in the Southern Department of the United States* (New York: University Publishing, 1869), 276.
[72] Lee, *The Campaign of 1781 in the Carolinas*, 177.

eminence, and put this part of the American line to flight: But no sooner was it done, than another line of the Americans presented itself to view . . . so as almost to encompass them. . . . The enemy's fire being repeated and continued . . . not only on the front but flank of the battalion . . . it was at last entirely broken. Fortunately . . . the Hessian regiment of Bose . . . was advancing in firm and compact order on the left of the guards, to attack the enemy. Lieutenant-colonel Norton . . . requested the Hessian lieutenant-colonel to wheel his regiment to the right and cover the guards, whilst the officers endeavoured to rally them. . . . The battalion, being again formed, instantly moved forward to join the Hessians. . . . No sooner had the guards and Hessians defeated the enemy in front, than they found it necessary to return and attack another body . . . in the rear; and in this manner were they obliged to traverse the same ground in various directions.[73]

Reverend Samuel Houston continued his vignette from the previous stand, relating his perspective of the battle within a battle:

Here we repulsed them again; and they a second time made us retreat back to our first ground; where we were deceived by a reinforcement of Hessians, whom we took for our own, and cried to them to see if they were our friends, and shouted Liberty! Liberty and advanced up till they let off some guns; then we fired sharply on them, and made them retreat a little. But presently the light horse [Tarleton's squadron] came on us, and not being defended by our own light horse, nor reinforced—though firing was long ceased in all other parts, we were obliged to run, and many were sore chased, and some cut down. We lost our major and captain then, the battle lasting two hours and twenty-five minutes.[74]

Analysis

1. In his postwar account Lieutenant Colonel Lee justified his departure by claiming he was merely repositioning his corps to protect Greene's left flank. By doing so, Lee left the attached Virginia and North Carolina troops unsupported against Tarleton's dragoons.
   a. Was Lee's decision in keeping with his specified and implied orders?
   b. Using the values and warrior ethos outlined in *Army Leadership*, ADRP 6-22, or *Marine Corps Operations* as a guide, further analyze Lee's decision. Does his decision agree with or deviate from the Army values? Does Lee's decision follow the enduring Marine Corps principle that Marines take care of their own?[75]
2. What are some ways that Lee could have coordinated the withdrawal with Colonel William Campbell, commander of the left flank riflemen?
3. Looking at the British perspective, how did Leslie's pursuit of Lee meet or deviate from Cornwallis's specified or implied command guidance?

Stand 10: The Third Line—Webster's Attack
Directions

Stand 10 is a brisk walk from the NPS "Expanding Battle" marker. Exercising caution, walk east along the paved vehicle trail (red route) approximately 300 meters to the parking lot at NPS marker 5.[76]

[73] Stedman, *The History of the Origin, Progress, and Termination of the American War*, vol. 2, 341–43.
[74] Graves, "Southern Campaigns American Revolution Pension Statements: Journal of Samuel Houston."
[75] *Army Leadership*, ADRP 6-22 (Washington, DC: Headquarters, U.S. Army, 2012), 3-1–3-4; and *Marine Corps Operations*, 1-5.
[76] The location of the marker does not correspond to the actual location of the third line fight as determined by modern NPS research. Refer to the adjacent NPS marker "Legend vs. Reality" for more details.

At the north corner of the parking lot by stop 5 is a paved foot trail, which passes close to an obelisk commemorating Lieutenant Colonel William Washington's cavalry regiment at Guilford Courthouse. Follow the paved trail to the northwest as it drops into a natural swalecut by a small branch of Richland Creek. If the vegetation is not too thick, this swale provides a distant overview of the third line battle position. To the northwest is a small white marker resembling a chess piece, which commemorates where British commander Lieutenant Colonel James Stuart fell in the final stages of the battle. The British line was to the northwest of the marker, while the approximate direction of the American third line, hidden from view in a belt of trees, was to the northeast. Resume walking as the trail drops into the swale and crosses the graveled New Garden Road. Guided by signs to stop 7, continue to follow the foot trail to the north as it parallels the tree line on your right (west). After a brief pause at a small, rectangular, white stone marker commemorating the fatal wounding of Virginia dragoon Captain Griffin Fauntleroy during one of Washington's attacks, resume walking northwest an additional 80 meters or so to the NPS interpretive panel titled "The British Perspective."

Orientation

This stand covers the opening phase of the third line fight, which lasted from approximately 1445–1500 that afternoon. Facing the "British Perspective" marker, orient the participants south. The point where the tree line ends at the west edge of the swale marks the approximate location where Lieutenant Colonel Webster led the *33d Foot* and supporting light infantry in the first attack on the Continental line. The apex of the Continental third line was about 300 meters due east of this point, hidden in the tree line. To the south-southeast at the bottom of the swale is the chess piece-like monument to Lieutenant Colonel Stuart, which marks the general location for the final skirmish of the battle. Adjacent to the stand location is a tall cylindrical monument, which was placed there in 1880 by the Guilford Courthouse Centennial Association, labeled on NPS maps as the "Regulars Marker."

Description

The Guilford Courthouse Centennial Association was responsible for the first purchase of land that was eventually incorporated as part of the NPS preserved battlefield. Back in 1880 when the association placed the "Regulars Marker," members believed the Continentals' main position ran along the edge of the open swale. However, later work by NPS researchers pinpointed the actual location of the Continental third line about 300 meters east of this location, on a ridgeline overlooking Richland Creek.[77] When describing the battlefield of 1781, the instructor must ensure that students understand that the visible heavy belt of trees along the east slope did not exist. Instead, this area was described as

> clear of wood, having only the usual growth of shrubbery, or small saplings, found in old fields. Alongside the road also, in the lane between the fields, there was a growth of saplings which overtopped the fences. . . . Below the eminence on which these troops are drawn up (Continentals), extends a plain intersected, at irregular intervals, by ravines or hollows . . . and terminating in a rising ground, at the edge of the wood which bounds the cleared land, and to the west of the road—this wood is about two hundred yards distant.[78]

After approximately 90 minutes of running battle, the first element of Webster's brigade, the *33d Foot*, with the *Guards Light Infantry* and Ansbach jäger companies in echelon, reached the west edge of the swale. After a hasty visual reconnaissance, Webster concluded he only faced a small visible contingent of Continentals and a battery of four Continental cannons emplaced behind "oldfields"

---

[77] Babits and Howard, *Long, Obstinate, and Bloody*, 238.
[78] Johnson, *Sketches of the Life and Correspondence of Nathanael Greene, Major General of the Armies of the United States, in the War of the Revolution*, vol. 2, 5.

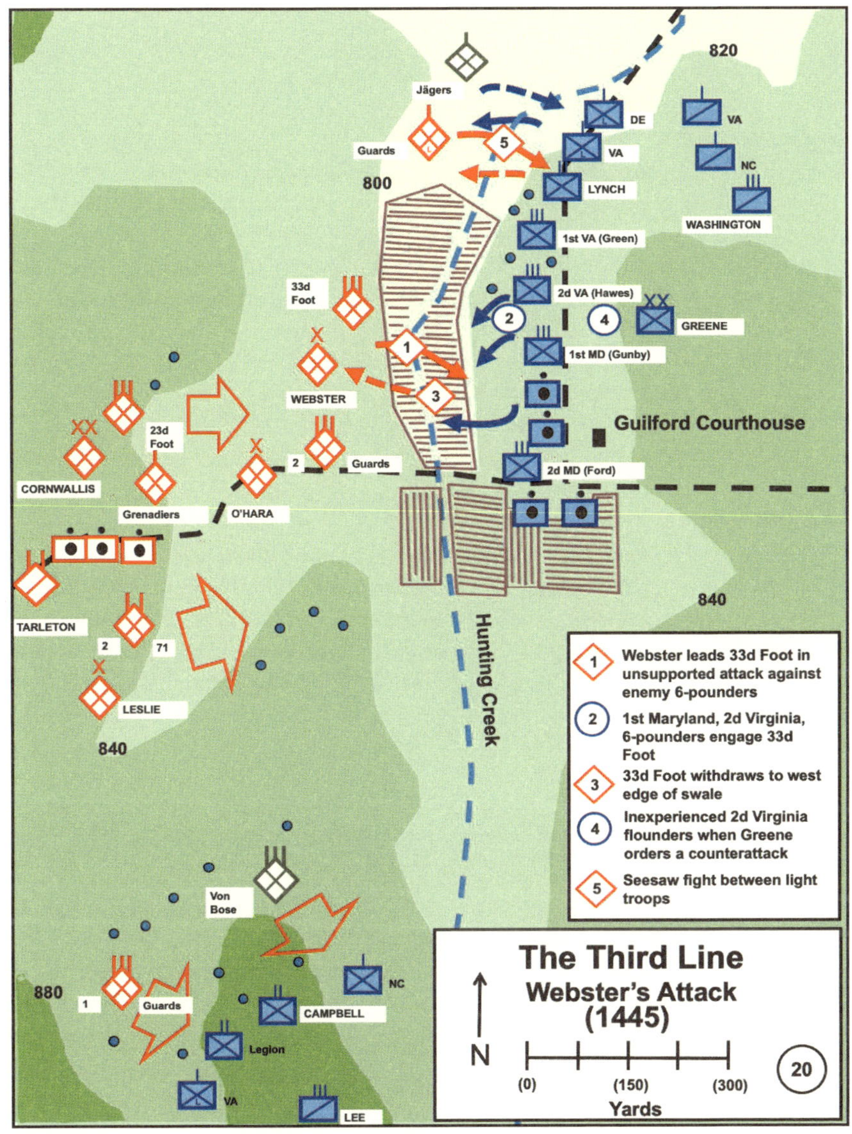

Map 20
Chuck Collins, Army University Press, and Harold Allen Skinner Jr.

overlooking the Salisbury Road. Webster apparently assumed he only faced Captain Robert Kirkwood's and Colonel Charles Lynch's independent Continental companies that had fought a rearguard action from the first militia line.

Instead, Webster faced two fresh Continental brigades of some 1,500 troops, supported by four 6-pounder cannons. Webster launched his attack with a 1:5 force ratio as his remaining regiments, the *23d Foot* and *2d Guards* battalion, were still fighting their way through the lethally resisting fragments of Lawson's Virginia brigade. The *33d Foot* was struck in rapid succession by volleys from the 1st Maryland and 2d Virginia Regiments and case shot from the Continental artillery. Stunned by the heavy fire, and the appearance of the previously undetected 2d Virginia, the *33d Foot* fell back to a hasty defensive position on the west side of the swale. Greene ordered a counterattack by the

2d Maryland and 1st Virginia, but the green regiments balked at crossing the kill zone in the swale, forcing Greene to call off the attack. During this phase of the battle, Webster was seriously wounded but refused to relinquish command.

Shortly afterward, the *2d Guards* battalion reached the west edge of the swale via the Great Salisbury Road. Brigadier General Charles O'Hara repeated Webster's error by leading the *2d Guards* in a hasty charge toward the clearly visible Continentals. Still disordered from their failed attack, the *33d Foot* could only support by fire from the edge of the swale. However, O'Hara's attack fared better than Webster's by hitting the 2d Maryland with enfilade fire as the unit was repositioning to face the threat. The 2d Maryland, a new regiment with green recruits and new officers led by Lieutenant Colonel Benjamin Ford, fell apart during the stressful maneuver, "unaccountably, the second regiment gave way, abandoning to the enemy the two field pieces."[79]

Vignettes

Commissary General Stedman recounted Webster's running fight to the third line.

> And lieutenant-colonel Webster, finding the left of the thirty-third regiment exposed to a heavy fire from the right wing of the enemy, which greatly out-flanked him, changed its front to the left; and the ground . . . was immediately occupied by general O'Hara. . . . Webster, moving to the left with the thirty-third regiment, supported by the light-infantry of the guards, and the yagers, routed and put to flight the right wing of the enemy, and in his progress, after two severe struggles, gained the right of the continentals; but the superiority of their numbers, and the weight of their fire, obliged him, separated as he was from the rest of the British line, to re-cross a ravine, and occupy an advantageous position on the opposite bank, until he could hear of the progress of the king's troops on the right.[80]

Lieutenant Colonel John Eager Howard of the 1st Maryland succinctly described the collapse of the 2d Maryland.

> The [Second Battalion of] guards after they had defeated Genl. [Edward] Stevens pushed into the cleared ground and run at the 2d [Maryland] regiment, which immediately gave way, owing I believe to the want of officers and having so many new recruits. The guards pursued them into our rear where they took two pieces of artillery.[81]

Analysis

1. Consider Webster's decision to attack in light of *Offense and Defense*'s characteristics of the offense: audacity, concentration, surprise, and tempo, or *Marine Corps Operations*'s tactical tenets for conducting expeditionary operations: achieving a decision, gaining advantage, tempo, adapting, and exploiting success and finishing.[82] Which characteristics did Webster follow or fail to follow? What was the outcome?
2. How could Greene and his subordinate commanders have better prepared the new 2d Maryland and 1st Virginia for the battle?
3. What about O'Hara's decision to attack without making coordination with Webster? What were the pros and cons?

---

[79] Col Otho Holland Williams, as quoted in Babits and Howard, *Long, Obstinate and Bloody*, 147.
[80] Stedman, *The History of the Origin, Progress, and Termination of the American War*, vol. 2, 339–40.
[81] John Eager Howard to John Marshall, undated, "Key Original Source Outline for the Battle of Guilford Courthouse," hereafter Howard to Marshall undated.
[82] *Offense and Defense*, 3-1–3-3; and *Marine Corps Operations*, 3-26–3-27.

Stand 11: The Third Line—
The 2d Guards and 1st Maryland Engagement

Directions

Stand 11 is in close proximity to the NPS "British Attack" marker. Retrace your steps south to the white Fauntleroy stone marker at stand 10, then turn right (west) and walk about 10 meters along the New Garden Road to the "Death of Stuart" NPS interpretive panel.

Orientation

Once at the "Death of Stuart" interpretive panel, face due west where the Salisbury Road descends into the open swale, the location where Brigadier General Charles O'Hara led his brigade attack on the American left flank. To the northeast is the small white chess piece commemorating British Lieutenant Colonel James Stuart. Beyond the Stuart marker, at the edge of the swale was the location where Webster's brigade regrouped after its repulse by the American line. Chronologically, this stand covers about 15 minutes of battle, from 1500 to 1515 hours on the afternoon of 15 March 1781.

Description

Both this stand and stand 12 discuss the major engagement between two of the best units in either army, the British *2d Guards* battalion and the Continental 1st Maryland Regiment. To avoid further backtracking, stand 11 discusses the British perspective of this engagement, while stand 12 focuses on the American perspective of the same engagement.

To recap, Webster's *33d Foot* and its supporting light infantry companies were hit by concentrated volley fire from the 1st Maryland and 2d Virginia, forcing the British to recoil to the northwest side of the swale. Shortly afterward, O'Hara led the *2d Guards* in an equally hasty attack, which succeeded in completely breaking the 2d Maryland and capturing a Continental 6-pounder gun section.

The impetus of O'Hara's attack quickly dissipated when the *2d Guards* battalion was struck on the left flank by synchronized volleys from the undetected 1st Maryland. As O'Hara struggled to reorient the *2d Guards* toward the threat, Lieutenant Colonel Washington's dragoon squadron hit the British flank and rear. O'Hara had little choice but to attempt a fighting withdrawal toward Webster's position. Colonel John Gunby of the 1st Maryland refused to break contact and a bloody close-range engagement rolled slowly westward into the swale. Casualties among the British leaders were heavy, with O'Hara seriously wounded and battalion commander Lieutenant Colonel James Stuart struck dead in the middle of the hand-to-hand melee. At the height of the infantry scrum, Cornwallis arrived on the west edge of the swale, leading the *Royal Artillery* 3-pounders, Tarleton's dragoons, and the *71st Foot*. The British 3-pounder crews swung into position, firing a blast of grapeshot to drive back the Americans, allowing O'Hara to regroup his battalion at the edge of the woodline. Cornwallis's effective use of grapeshot apparently caused casualties among the guardsmen, giving rise to the assertion that Cornwallis cold-bloodedly ordered indiscriminate firing into the melee.[83] Faced with the threat of fresh British reinforcements, the 1st Maryland and Washington's dragoons withdrew about 300 meters to the vicinity of Reedy Fork Road.

Vignettes

An excerpt from Cornwallis's post-battle report describes the engagement in the swale.

> The 2d battalion of Guards first gained the cleared ground near Guildford court house,
> and found a corps of continental infantry, much superior in number, formed in the open

---

[83] Babits and Howard, *Long, Obstinate, and Bloody*, 161–62, thoroughly debunks the theory of Cornwallis deliberately firing into his own men.

field on the left of the road. Glowing with impatience to signalize themselves, they instantly attacked and defeated them, taking two six-pounders; but, pursuing into the wood with too much ardour, were thrown into confusion by a heavy fire, and immediately charged and driven back into the field by Colonel [William] Washington's dragoons, with the loss of the six-pounders they had taken. The enemy's cavalry was soon repulsed by a well-directed fire from two three-pounders just brought up by Lieutenant [John] Macleod, and by the appearance of the grenadiers of the guards, and of the 71st regiment, which, having been impeded by some deep ravines, were now coming out of the wood on the right of the guards, opposite to the court house.[84]

Close-in combat during the Revolutionary War was often a bloody, brutal affair, as was graphically depicted in a later account of the duel between Captain John Smith of the 1st Maryland Regiment and Lieutenant Colonel Stuart of the *2d Guards* battalion.

[Captain] Smith and his men were in a throng, killing the Guards and Grenadiers like so many Furies, Col. Stewart, seeing the mischief Smith was doing, made up to him through the crowd, dust, and smoke, and made a violent lunge at him with his small sword. The first that Smith saw was the shining metal . . . he only had time to lean a little to the right, and lift up his left arm . . . it would have been through his body but for the haste of the colonel and happening to set his foot on the arm of a man Smith had just cut down . . . his unsteady step, his violent lunge and missing his aim brought him with one knee upon the dead man . . . Smith had no alternative but to wheel round to the right and give Stewart a back handed blow over or across the head on which he fell; [Stuart's] orderly sergeant attacked Smith, but Smith's sergeant dispatched him; a 2nd attacked him Smith hewed him down, a 3rd behind him threw down a cartridge and shot him in the back of the head, Smith now fell among the slain but was taken up by his men and brought off, it was found to be only a buckshot lodged against the skull and had only stunned him.[85]

Analysis

1. Compare and contrast the British and American conformance or deviation from the principle of mass and maneuver. What was the impact on the fight at the third line?
2. During the fight at the third line, both sides were surprised by the unexpected appearance of enemy forces. Using the security tactical enabling task as outlined in *Offense and Defense*'s "Security Operations" section (paragraph 5-11) or *Marine Corps Operations*'s "Reconnaissance and Security Operations" section as a guide, what were some things both sides could have done to avoid surprise?[86]
3. Analyze the pros and cons of O'Hara's decision to execute a hasty attack versus deliberate attack? What could O'Hara have done to improve the odds of success?

Stand 12: The Third Line—Continental Counterattack
Directions

From the "Death of Stuart" interpretive panel, turn north and follow the paved trail to the T inter-

---

[84] Cornwallis letter, 17 March 1781.

[85] Babits and Howard, *Long, Obstinate, and Bloody*, 159.

[86] "Security operations are those operations undertaken by a commander to provide early and accurate warning of enemy operations, to provide the force being protected with time and maneuver space within which to react to the enemy, and to develop the situation to allow the commander to effectively use the protected force." *Offense and Defense*, 5-3; and *Marine Corps Operations*, 6-4.

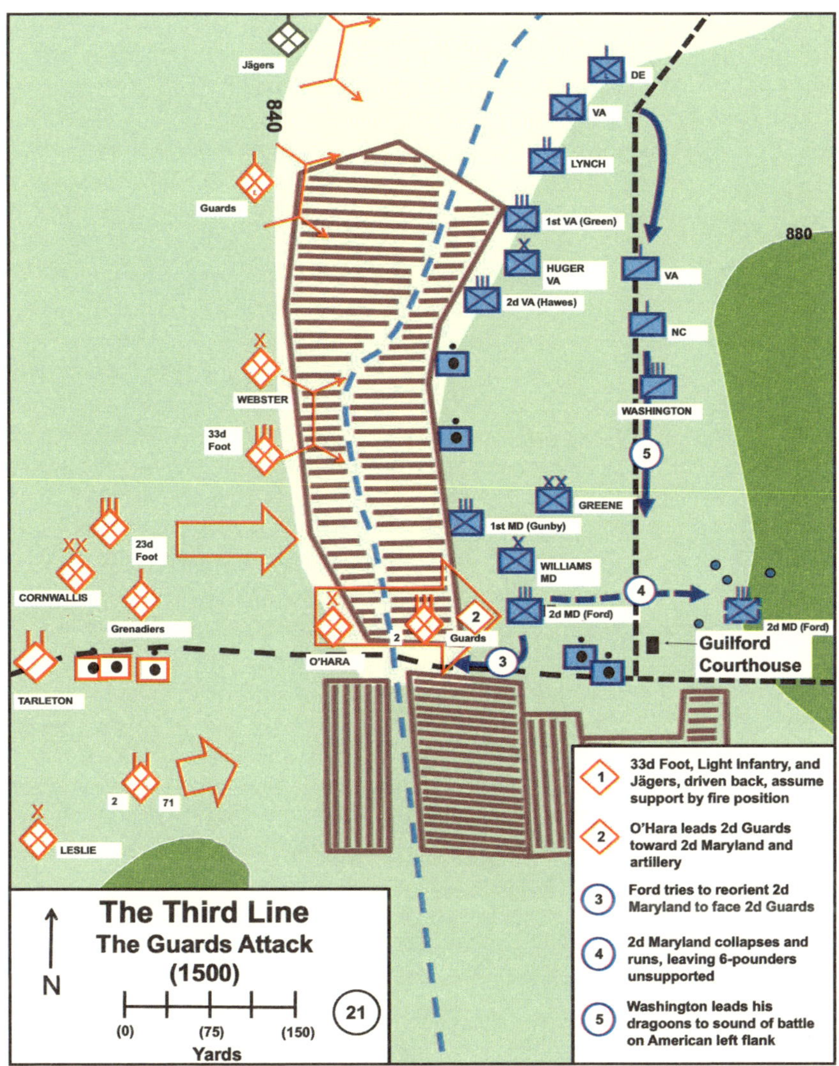

Map 21
Chuck Collins, Army University Press, and Harold Allen Skinner Jr.

section near the Fauntleroy marker at stand 10. Following signs for the NPS stop 6 marker, turn right (east) and walk along the trail as it eventually drops down to the level of Richland Creek. Continue along the trail, crossing bridges no. 4 and 3, until the trail turns left to climb up the shallow ridgeline dominating the east bank of Richland Creek. After a walk of roughly 300 meters, you will arrive at a sharp bend in the trail dominated by two reproduction Continental cannons. Stop at the NPS interpretive panel titled "American Artillery."

Orientation

The position of stand 12 overlooks the approximate starting position of the 1st Maryland Regiment before its attack on the *2d Guards* battalion. Orient first to face the "American Artillery" interpretive marker, then point out the paved trace of the New Garden Road to the south. Pointing back down the paved trail to the west indicates the general avenue of approach for the *2d Guards* battalion, beyond which was the approximate location for the final engagement at the third line. On the open ground to the south, running parallel to the Salisbury Road, was the position of the 2d Maryland, while the 1st Maryland was oriented facing west, with its line running almost due north of this location. To the northeast about 13 miles (21 km) was the location of Speedwell's Iron Works, General Greene's predesignated rally point after the battle.

Description

This stand discusses the American perspective of the climactic engagement in the swale. Greene's third line of defense was here, four Continental regiments supported by 6-pounder guns. To the south and east, positioned parallel to the New Garden Road, was the green 2d Maryland supporting the two-gun Continental artillery section. The 1st Maryland was concealed in the woods just north of this position, while Brigadier General Isaac Huger's Virginia brigade, and the remaining 6-pounder guns were to the north and east of this point. Huger's tactical task was to defend the right flank thus preventing the British from cutting the withdrawal route to Speedwell Iron Works.

As noted earlier, the poorly positioned 2d Maryland invited attack by O'Hara's *2d Guards* battalion. Lieutenant Colonel Benjamin Ford's attempt to reorient the 2d Maryland disordered the inexperienced soldiers. Lieutenant Colonel Otho Williams, the Maryland-Delaware brigade commander, made matters worse by ordering the companies to fall backward and regroup. The 2d Maryland barely managed to get into line and fire a ragged volley before the *2d Guards* battalion had closed into bayonet range. Faced with a wall of bayonets, the green Maryland troops broke and ran toward the courthouse. A lone North Carolina Continental company, commanded by Captain Edward Yarborough, and the Continental gunners skirmished briefly before abandoning the two 6-pounders to the guardsmen. The ensuing pursuit of the fleeing Continental soldiers left the understrength *Guards* battalion overextended and vulnerable to counterattack.

Lieutenant Colonel John Eager Howard, second in command of the 1st Maryland, saw the opportunity and "rode to Col. John Gunby and gave him the information. He did not hesitate to order the regiment to face about."[87] Firing by division to gain fire superiority, the veteran 1st Maryland flayed the vulnerable flank of the unsupported *2d Guards* battalion with volleys of "buck and ball." As the *Guards'* officers and noncommissioned officers fought to keep their soldiers in formation in the face of a Continental bayonet charge, Lieutenant Colonel Washington's dragoons hit the *Guards'* right flank and rear in a perfectly timed attack. Hit in front and flank, the *2d Guards* battalion recoiled and for several minutes a brutal close ranged skirmish took place between the three units.

Just as victory seemed assured for the Continentals, a blast of British grapeshot announced the arrival of Cornwallis and additional British reinforcements. Several Continental fighters were killed and wounded, prompting both Washington and Gunby to break off the pursuit of O'Hara and withdraw toward Reedy Fork Road. The Continental cannon crews abandoned their heavy 6-pounders as several horses had been killed or wounded. Under the cover of the Huger's Virginia brigade, the Maryland Brigade regrouped on the Reedy Fork Road to delay the expected British advance.

Vignettes

Lieutenant Colonel Williams described the American third line deployment.

> The 1st regiment under Gunby was formed in a hollow in the wood to the right of the cleared ground about the Courthouse. The Virginia Brigade under Genl. Huger were to

---

[87] Howard to Marshall undated.

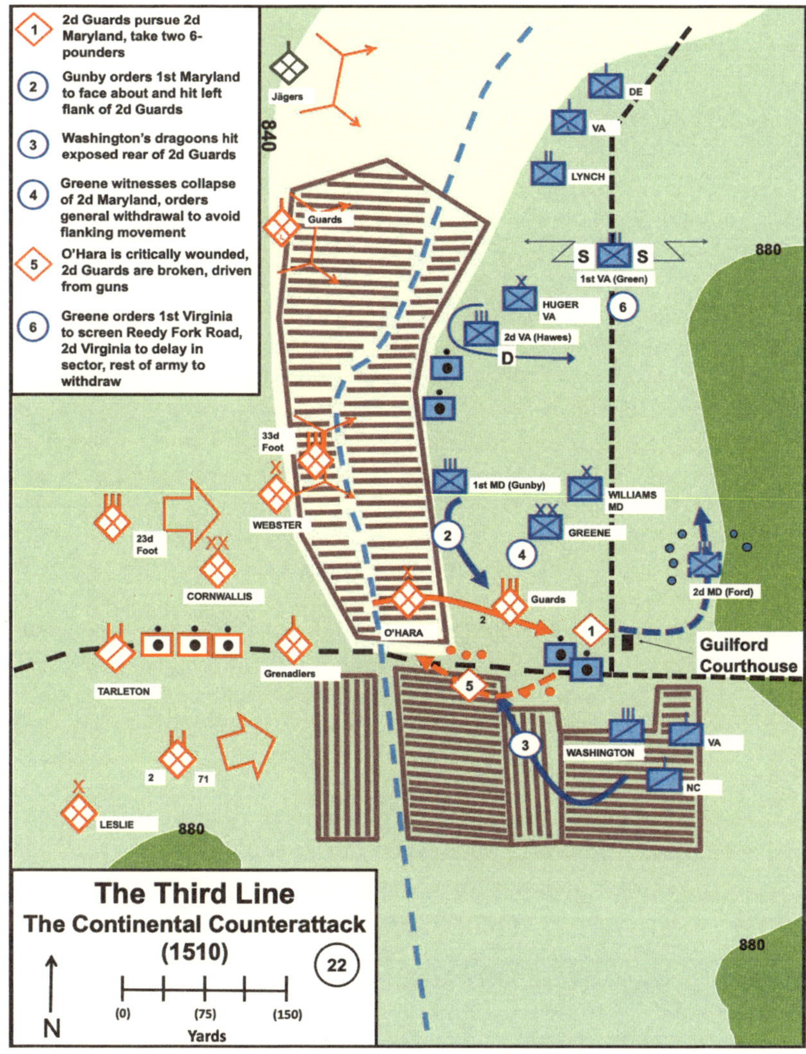

Map 22
Chuck Collins, Army University Press, and Harold Allen Skinner Jr.

our right. The second regiment [2d Maryland] was at some distance to the left of the first, in the cleared ground with its left flank thrown back so as to form a line almost at right angles with the 1st regt.[88]

Lieutenant Colonel Howard described the counterattack of the 1st Maryland.

Capt. [Jonathan] Gibson . . . rode to me and informed me that a party of the enemy . . . were pushing through the cleared ground and into our rear, and if we would face about and charge them, we might take them. We had been for some time engaged with a part of Webster's Brigade, though not hard pressed, and at that moment their fire had slackened. I rode to [Col. John] Gunby and gave him the information. He did not hesitate to order the

<hr>

[88] "Letter Answering Questions on Cowpens, Guilford Courthouse, and Hobkirk's Hill," in John H. Beakes Jr., *Otho Holland Williams in the American Revolution* (Charleston, SC: Nautical and Aviation Publishing, 2015), loc. 4298 of 9123, Kindle.

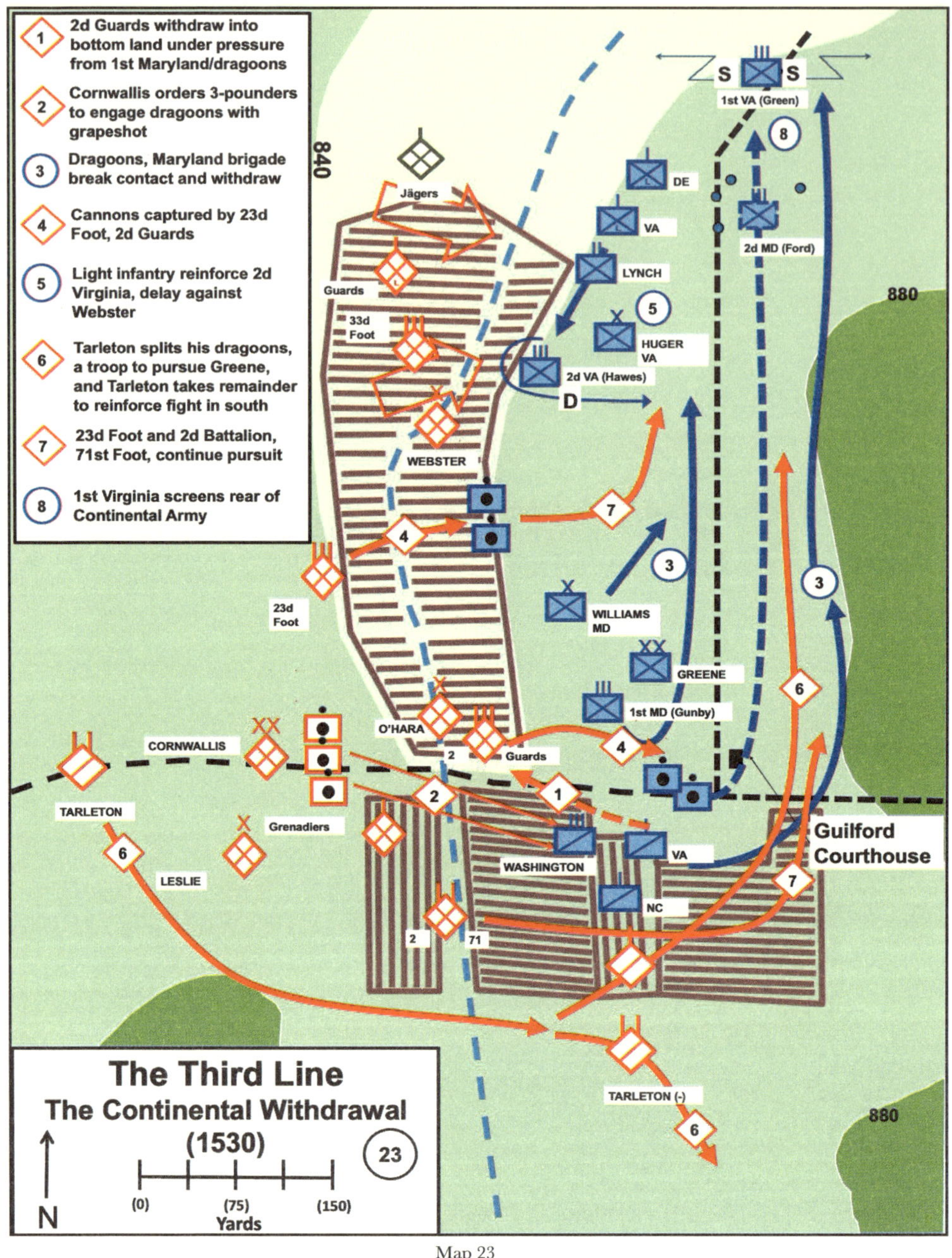

Map 23
Chuck Collins, Army University Press, and Harold Allen Skinner Jr.

regiment to face about and we were immediately engaged with the Guards. Our men gave them some well-directed fires and we then advanced and continued firing.[89]

Sergeant Major Seymour of Kirkwood's Continental Light Infantry described the impact of the Continental dragoon attack on O'Hara's brigade.

---

[89] Howard to Marshal undated.

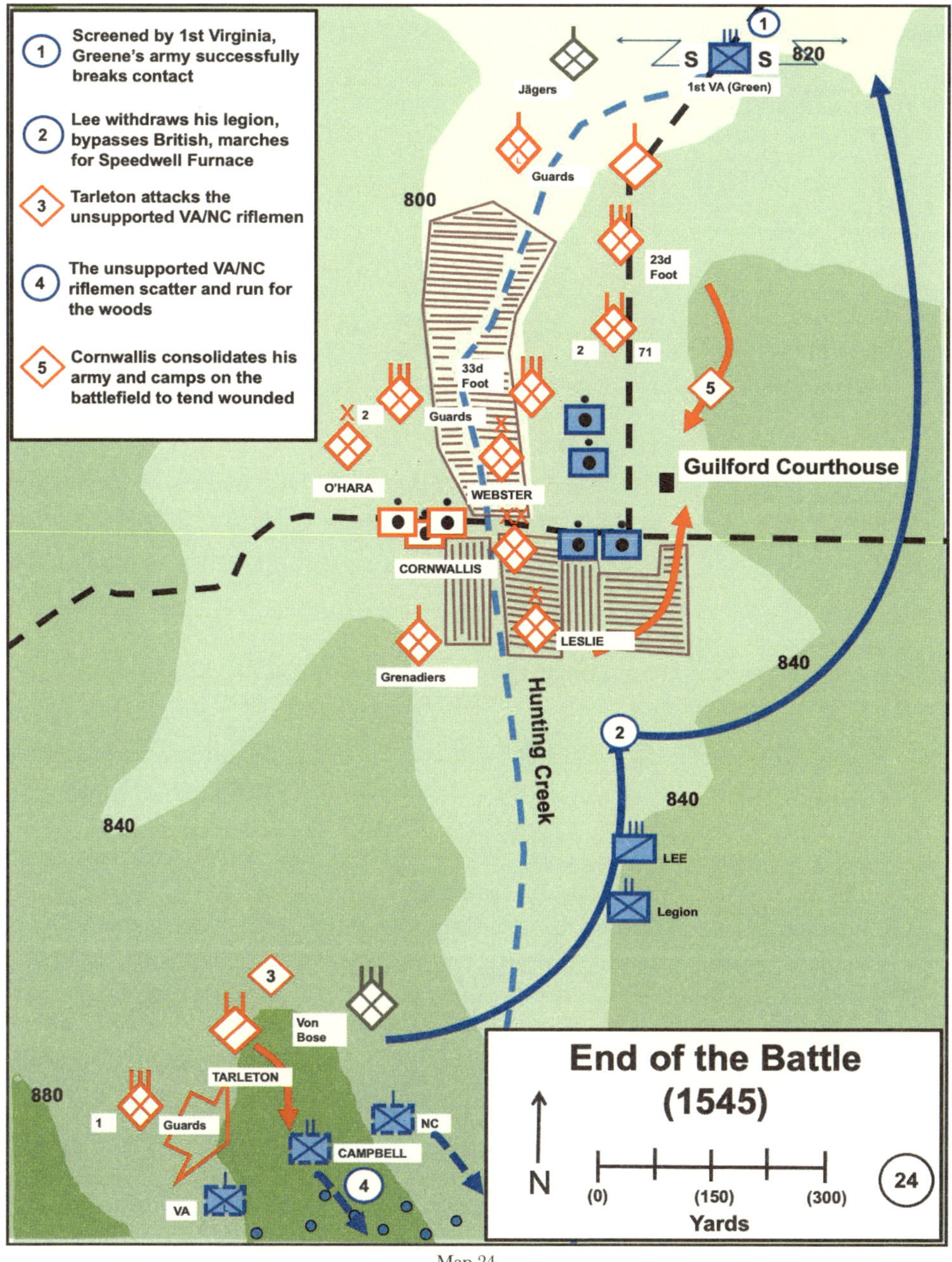

Map 24
Chuck Collins, Army University Press, and Harold Allen Skinner Jr.

Colonel Washington, with his cavalry, in this action deserved the highest praise, who meeting with the Third Regiment of Foot Guards, and charged them so furiously that they either killed or wounded almost every man in the regiment, charging through them and breaking their ranks three or four times.[90]

---

[90] Seymour, "A Journal of the Southern Expedition, 1780–1783," 378.

After his withdrawal from the first line, Private Nathan Slade watched the third line fight from a good vantage point at the courthouse.

> This conflict between the brigade of Guards and the first regiment of Marylanders was most terrific, for they fired at the same instant, and they appeared so near that the blazes from their guns seemed to meet.[91]

Analysis

1. The breaking of the 2d Maryland Regiment highlights the risks associated with sending new soldiers into combat. What were some things General Greene and his subordinates could have done to improve the combat readiness of the unit and/or mitigate the risk?

2. The initial position of the 2d Maryland left the unit vulnerable to flank attack. In light of Greene's commander's intent and mission, what were the advantages and disadvantages of the position? What, if anything, could Greene have done differently?

3. Using the same line of reasoning, consider the positioning of the Continental 6-pounders. What could Greene have done to improve their effectiveness while decreasing their vulnerability to infantry assault?

4. The quick reaction of the 1st Maryland Regiment and Washington's dragoons indicate a high level of trust and tactical competence. What are some possible reasons?

5. In accounts after the battle, Cornwallis was often accused of firing on his own troops to force the Continentals to break their pursuit. What modern parallels or lessons can be drawn from the incident?

Stand 13: The Third Line—The Continentals' Withdrawal
Directions
From the artillery display, walk about 30 meters to the southeast until the trail intersects the historic New Garden Road and stop at the NPS interpretive marker titled "Third Line Trail."

Orientation
This stand will discuss how Generals Greene and Cornwallis responded to the swirling fight between the 1st Maryland and the *2d Guards* battalion, downslope in the swale. After orienting students to the trace of the New Garden Road, point northwest toward stand 12, the general vicinity of the fight between the *2d Guards* battalion and the 1st Maryland around the Continental cannons. Then, point to the north to indicate the general location of the Virginia brigade and northeast for the general direction of the Reedy Fork Road. Last, point out the difference in terrain between the creek bed and the probable trace of the Continental line along the military crest.

Description
Up to the moment when the 2d Maryland Regiment folded under pressure, General Greene had felt that he had the battle well in hand. When the 2d Maryland collapsed, Greene was nearly killed by the *2d Guards* when he attempted to rally the troops. Shaken by the experience, Greene naturally concluded that the Continental defense was no longer tenable and the route of withdrawal via the Reedy Fork Road was threatened by the *2d Guards*. As a consequence, Greene transitioned from an area defense to a delaying operation, continuing to inflict damage on the British, while seeking to break contact.

---

[91] Pension application of Nathan Slade, "Key Original Source Outline for the Battle of Guilford Courthouse."

First, Greene ordered Huger's unbloodied Virginia Continental brigade to screen the withdrawal of the 1st Maryland and Continental dragoons from the creek vale. Huger placed Colonel John Green's 1st Virginia Regiment to protect the Reedy Fork Road, while Lieutenant Colonel Samuel Hawes's 2d Virginia Regiment deployed to delay the British advance from west and south. Hawes's fresh troops skirmished to delay the advance of the *2d Guards* and *33d Foot* as the *23d* and *71st Foot* advanced to gain contact. Brigadier General Huger was wounded in this phase of the battle, which ended as Hawes's 2d Virginia broke contact and leapfrogged past Green's 1st Virginia, which assumed the delay mission. Dissuaded by the bayonets of the 1st Virginia, Cornwallis ordered his units to regroup around the captured cannons, thus ending the main battle around 1530.

Vignettes

Lieutenant Colonel Howard described the withdrawal of the 1st Maryland.

> After passing through the guards . . . I found myself in the cleared ground, and saw the 71st regt. near the court house and other columns of the enemy appearing in different directions, Washington's horse having gone off, I found it necessary to retire, which I did leisurely, but many of the guards who were laying on the ground and who we supposed were wounded, got up and fired at us as we retired.[92]

General Greene recounted his decision to withdraw.

> They having broken the 2d Maryland Regiment, and turned our left flank, got into the rear of the Virginia brigade (of 18-month regiments); and appearing to be gaining on our right, which would have encircled the whole of the Continental troops, I thought it most advisable to order a retreat. About this time Lieutenant colonel Washington made a charge with the horse upon a part of the brigade of guards; and the first regiment of Marylanders, commanded by Colonel Gunby, and seconded by Lt. Col. Howard, followed the horse with their bayonets; near the whole of the party fell a sacrifice.[93]

Analysis

1. Put yourself in Greene's boots. As the commander, what is more important to your overall operational goal: possession of the ground; retention of major combat systems (cannons); preservation of combat power; or destruction of the enemy force?
2. What other factors (e.g., physical, mental, and moral) could have affected Greene's decision-making at this stage in the battle?
3. What of Cornwallis? At this point in the battle, what was more important to the outcome of the operational goals: possession of the ground; capture of key weapon systems; destruction of the enemy force; or preservation of combat power? Instead, what did Cornwallis decide?

Stand 14: The Butcher's Bill
Directions

From the "Third Line Trail" marker, follow the gravel bed of the New Garden Road east (left) about 109 yards (100 meters) to the NPS "Backcountry Courthouse" interpretive marker.

---

[92] Howard to Marshall undated.

[93] Nathanael Greene to Samuel Huntington, President of the Continental Congress, 16 March 1781, "Key Original Source Outline for the Battle of Guilford Courthouse."

Orientation

This stand overlooks the approximate intersection of the Reedy Fork Road with the New Garden Road. Reedy Fork Road ran northeast from the courthouse toward the designated American rally point at Speedwell Ironworks. Point out the trace of the New Garden Road running east to west, then the approximate location of Guilford Courthouse on a hillock to the northeast. Finally, point back to the west to indicate the original location of the Continental third line.

Description

Concerned with the safety of his irreplaceable Continentals, Greene designated Speedwell Ironworks as a fallback position, around 13 miles (21 km) to the northeast of the courthouse. The American main body broke contact with the British around 1530 and marched wet but unopposed to the Ironworks, while the tired British troops camped on the battlefield. Little food was available, and the unwounded British soldiers watched many of their wounded die of exposure in a cold, driving rain.[94]

Greene's army was equally tired and wet and was much smaller in the morning on account of the wholesale desertion of the Carolina militia. Nevertheless, spirits in the Continental ranks were high because the troops believed they had bested the redcoats in battle. At the Speedwell Ironworks, a picket line and defensive works was established, wounded and stragglers collected, and the unwounded men regrouped with their units. Greene cannily boosted morale with a timely issue of two days' rations and a gill of rum from the Continental quartermaster stores and the issuance of an order praising the conduct of the Continental and Virginia troops. Despite having to abandon his wounded and artillery on the battlefield, Greene believed his army had inflicted disproportionate damage on Cornwallis's army. With his corps of Continentals intact, Greene was satisfied to leave Cornwallis alone and await developments.

Out of some 4,400 Continentals and militia, Greene's army suffered 79 dead, 184 wounded, and 1,046 missing—mostly North Carolina militia who deserted the fight for home. Discounting the handful of Americans taken prisoner, Greene's army suffered a 7-percent casualty rate. By contrast, Cornwallis's smaller army of 1,924 men suffered proportionally heavier casualties: 93 dead, 413 wounded, and 26 missing—a crippling 28-percent casualty rate.[95]

Before the battle, neither army made a systematic plan to handle their wounded and dead. Afterward, details of British soldiers, assisted by local Quakers and Moravians, collected the wounded to field hospitals set up in civilian houses, including the Hoskins homestead. Cornwallis and Greene quickly negotiated an agreement for the Americans to provide regimental surgeons and medical supplies. Throughout the next day after the battle, American and British surgeons worked together in treating the injured. Meanwhile, British soldiers collected the dead, interring many on the battlefield, while those that died in the hospitals were buried in common graves.

After two days of rest and treatment, Cornwallis rejoined the field trains at New Garden Meeting House. While his troops rested and refitted, Cornwallis announced his great victory over the rebels and called on Loyalists to join his army. The Tories in the area sized up the condition of Cornwallis's troops and shrewdly decided not to risk retribution from the Patriots. After a couple of days, Cornwallis realized no help was coming. He wryly noted afterward, "I could not get one hundred men in all the regulator's country to stay with us even as militia."[96] Consequently, Cornwallis marched to the Loyalist enclave at Cross Creek, North Carolina, where he hoped to resupply and open his line of communications with the British depot at Wilmington, North Carolina. In a final failure of British

[94] Babits and Howard, *Long, Obstinate, and Bloody*, 172.
[95] Babits and Howard, *Long, Obstinate, and Bloody*, 173–75.
[96] Ross, *Correspondence of Charles, First Marquis Cornwallis*, 84; and Buchanan, *The Road to Guilford Courthouse*, 382.

intelligence, Cornwallis found Cape Fear River between Cross Creek and Wilmington unsuited for use as a supply line. With that final blow to his campaign plan, Cornwallis dejectedly withdrew to Wilmington.[97]

Greene used his corps of observation to monitor Cornwallis's dispositions, making no attempt to engage until Cornwallis withdrew toward Wilmington. British march discipline was excellent, with only a handful of stragglers captured by the Americans before the British Army reached safety at Wilmington. Greene lingered in the area until early April, when intelligence confirmed Cornwallis was marching toward Virginia, thus permanently ceding the initiative in the Carolinas. Greene then transitioned to a "war of the posts," the reduction of British bases in South Carolina.[98] Greene's attempt to besiege Camden ended on 25 April when Lieutenant Colonel Francis Rawdon-Hastings attacked Greene's camp near Hobkirk's Hill. Greene's men successfully repulsed the attack, but a failed counterattack so disordered the Americans that Greene was compelled to lift the siege. Despite his loss, Greene's presence in South Carolina led Rawdon-Hastings to evacuate the isolated Camden garrison on 10 May. The pattern of American tactical loss followed by strategic gain repeated itself at the siege of Ninety Six in June and the Battle of Eutaw Springs in September. After Eutaw Springs, the British Army withdrew into the vicinity of Charleston, where it remained, impotent, until evacuating a year later on 14 December 1782.[99]

As evident by his offensive into Virginia, Cornwallis had abandoned the notion of pacifying the Carolinas, believing the British Army incapable of pacifying the region without retaking Virginia.[100] After refitting his command and informing Lord Germain (but not Lieutenant General Henry Clinton) of the altered campaign plan, Cornwallis marched from Wilmington on 25 April toward Virginia, permanently ceding the strategic initiative in the Carolinas to General Greene. Instead of reducing Virginia, Cornwallis's army fought a series of inconclusive battles before entrenching itself at the port of Yorktown, Virginia, to await reinforcements. There, Cornwallis was trapped and besieged by a joint Franco-American task force and ultimately forced to surrender his army at Yorktown on 19 October 1781.

Vignettes

Commissary General Stedman described the aftermath of the Guilford Courthouse battle.

> The night of the day on which the action happened was remarkable for its darkness, accompanied with rain, which fell in torrents. Near fifty of the wounded, it is said, sinking under their aggravated miseries, expired before the morning. The cries of the wounded and dying who remained on the field of action during the night exceed all description. Such a complicated scene of horror and distress, it is hoped, for the sake of humanity, rarely occurs, even in a military life.[101]

General Greene's after action report of the Guilford Courthouse operation is as follows:

> Here has been the field for the exercise of genius, and an opportunity to practice all the great and little arts of war. Fortunately, we have blundered through without meeting with any capital misfortune. On the 11th of this month I formed a junction, at the High Rock Ford, with a considerable body of Virginia and North Carolina militia, and with a Virginia

[97] Buchanan, *The Road to Guilford Courthouse*, 383.
[98] George Washington to John Hancock, 8 September 1776, *Founders Online*, National Archives and Records Administration, accessed 22 June 2020.
[99] Babits and Howard, *Long, Obstinate, and Bloody*, 181–85.
[100] "Letter from General Cornwallis to Major General Phillips, 10 April March 1781," in Thomas Goss, ed., *Primary Source Reference Book for the 1781 Guilford Courthouse Campaign* (West Point, NY: Department of History, U.S. Military Academy, 1998), 7.
[101] Stedman, *The History of the Origin, Progress, and Termination of the American War*, vol. 2, 346.

regiment of eighteen months' men. Our force being now much more considerable than it had been, and upon a more permanent footing, I . . . [gave] the enemy battle without loss of time, and made the necessary dispositions accordingly. The battle was fought at or near Guilford Court-House. . . . The battle was long, obstinate, and bloody. We were obliged to give up the ground, and lost our artillery, but the enemy have been so soundly beaten that they dare not move towards us since the action, notwithstanding we lay within ten miles of him for two days. Except the ground and the artillery, they have gained no advantage; on the contrary, they are little short of being ruined. The enemy's loss in killed and wounded cannot be less than between six and seven hundred, perhaps more. Victory was long doubtful and had the North Carolina militia done their duty, it was certain. They had the most advantageous position I ever saw, and left it without making scarcely the shadow of opposition. Their General and Field Officers exerted themselves, but the men would not stand. Many threw away their arms, and fled with the utmost precipitation, even before a gun was fired at them. The Virginia militia behaved nobly, and annoyed the enemy greatly. The horse, at different times in the course of the day, performed wonders. Indeed, the horse is our great safeguard, and without them the militia could not keep the field in this country. . . . I am happy in the confidence of the army, and, though unfortunate, I lose none of their esteem. Never did an army labour under so many disadvantages as this; but the fortitude and patience of the officers and soldiery rise superior to all difficulties. . . . Our army is in good spirits, but the militia are leaving us in great numbers to return home to kiss their wives and sweethearts.[102]

Lieutenant General Cornwallis's follow-on report to Lord Germain concerning the outcome of the Guilford Courthouse battle read:

His Majesty's troops under my command obtained a signal victory on the 15th instant over the rebel army commanded by General Greene. . . . I had encamped on the 13th instant at the Quaker Meeting between the forks of the Deep River. On the 14th I received information that General Butler, with a body of North Carolina militia and the expected reinforcements from Virginia . . . had joined General Greene, and that the whole army, which was reported to amount to nine or ten thousand men, was marching to attack. . . . Being now persuaded that he had resolved to hazard an engagement, after detaching Lieut.-Colonel Hamilton with our wagons and baggage . . . I marched with the rest of the corps at daybreak on the morning of the 15th to meet the enemy. . . . About four miles from Guildford our advanced guard, commanded by Lieutenant-Colonel Tarleton, fell in with a corps of the enemy, consisting of Lee's legion . . . which he attacked. . . . and defeated it and, continuing our march, we found the rebel army posted on rising ground, about a mile and a half from the Court-house. The prisoners taken by Lieut.-Colonel Tarleton having been several days with the advanced corps, could give me no account of the enemy's order or position, and the country. Immediately between the head of the column and the enemy's line was a considerable plantation. . . . The wood . . . in the skirt of which the enemy's first line was formed, was about a mile in depth, the road then leading into an extensive space of cleared ground about Guilford Court-house. The woods on our right and left were reported to be impracticable for cannon; but, as [the

---

[102] "Letter from General Nathanael Greene to Joseph Reed, Adjutant General of the Army, 18 March 1781," in Goss, *Primary Source Reference Book for the 1781 Guilford Courthouse Campaign*, 42.

woods] on our right appeared the most open, resolved to attack the left wing of the enemy
. . . I ordered Lieutenant Macleod to bring forward the guns and cannonade their center.
. . . On the right, the regiment of Bose and the 71st regiment led by Major-General Leslie,
and supported by the first battalion of Guards; on the left, the 23rd and 33rd regiments,
led by Lieut.-Colonel Webster, and supported by the Grenadiers and second battalion of
Guards commanded by Brigadier-General O'Hara; the Yagers, and light infantry of the
Guards, remained in the wood on the left of the guns, and the cavalry in the road, ready to
act as circumstances might require . . . the action began about half-an-hour past one in the
afternoon. . . . Major-General Leslie after been obliged, by the great extent of the enemy's
line, to bring up the first battalion of Guards to the right of the regiment of Bose, soon
defeated everything before him. Lieut.-Colonel Webster, having joined the left of Major-
General Leslie's division, was no less successful in his front, when, on finding that the left of
the 23rd was exposed to a heavy fire from the right wing of the enemy, he changed his front
to the left, and being supported by the Yagers and light infantry of the Guards, attacked
and routed it, the grenadiers and second battalion of Guards moving forward to occupy
the ground left vacant by the movement of Lieut.-Colonel Webster. . . . The excessive
thickness of the woods rendered our bayonets of little use, and enabled the broken enemy
to make frequent stands with an irregular fire, which occasioned some loss . . . particularly
on our right, where the first battalion of the Guards and regiment of Bose were warmly
engaged in front, flank, and rear. . . . The second battalion of Guards first gained the
clear ground near Guilford Court-house, and found a corps of Continentals infantry. . . .
they instantly attacked and defeated them, taking two 6-pounders, but, pursuing into the
wood with too much ardor, were thrown into confusion by a heavy fire, and immediately
charged and driven back into the field by Colonel Washington's dragoons, the loss of the
6-pounders they had taken. The enemy's cavalry was soon repulsed by a well-directed fire
from two 3-pounders . . . and by the appearance of the grenadiers of the Guards, and of
the 71st regiment, which, having been impeded by some deep ravine, were now coming
out of the wood on the right of the Guards, opposite to the Court-house. By the spirited
exertions of Brigadier General O'Hara, though wounded, the second battalion of Guards
was soon rallied . . . the enemy were soon put to flight, and . . . all the artillery they had in
the field, were likewise taken. . . . The 23rd and 71st regiments, with part of the cavalry,
were ordered to pursue; the remainder of the cavalry was detached; with Lieut.-Colonel
Tarleton, to our right . . . where his appearance and spirited attack contributed much to a
speedy termination of the action. The militia . . . dispersed in the woods; the continentals
went off by the Reedy Fork, beyond which it was not in my power to follow them, as their
cavalry had suffered but little. Our troops were excessively fatigued by an action which
lasted an hour and a half, and our wounded . . . required immediate attention. The care
of our wounded, and the total want of provisions in an exhausted country, made it equally
impossible for me to follow. . . . From our observation . . . we did not doubt but the strength
of the enemy exceeded 7000 men. . . . I cannot ascertain the loss of the enemy, but it must
have been considerable; between 200 and 300 dead were left upon the field. Many of their
wounded that were able to move . . . escaped . . . the houses in a circle of six or eight miles
round us are full of others. Those that remained we have taken the best care of in our
power. . . . The conduct and actions of the officers and soldiers that composed this little
army will do more justice to their merit than I can by words. Their persevering intrepidity
in action, their invincible patience in the hardships and fatigues of a march of above 600

miles . . . without tents or covering against the climate, and often without provisions, will sufficiently manifest their ardent zeal for the honor and interests of their Sovereign and their country.[103]

Analysis

1. Neither commander seemed to have made a plan for handling casualties. Taking into account the technology of the day, what were some basic planning measures each command could have taken?

2. Consider the same question in today's terms. What are your legal and moral obligations in caring for the dead and wounded?

3. Critically consider Lieutenant General Cornwallis's leadership performance during the campaign. First, what were his strengths? Second, what were his weaknesses? What of his subordinate commanders?

4. Consider the same questions regarding Major General Nathanael Greene and of his subordinate commanders.

5. In letters after the battle, Greene's comments about the collapse of the North Carolina militia seem overblown and inaccurate. Consider why Greene would write in such a way. What did he stand to gain or lose? Can we draw some parallels with today's multicomponent (Active and Reserve) force?

6. Using *Offense and Defense*, paragraphs 1-37 thru 1-48, "Solving Tactical Problems," or *Marine Corps Operations*, chapter 6 "Major Operations and Campaigns," as a guide, analyze Greene's performance as an operational commander:[104]

   a. When and how did Greene gain the tactical initiative over Cornwallis?

   b. How did Greene control the tempo of operations during the campaign?

   c. How did Greene seem to use commander's visualization to arrive at tactical solutions?

7. What lessons can we derive from the experiences of the officers and soldiers?

This marks the end of the battlefield terrain walk. After allowing students an opportunity to use the nearby toilet and water facilities, the suggested location for the Part IV, Integration, event is beneath one of the large live oak trees at the northeast edge of the clearing. In the event of inclement weather, the classroom in the visitors center is recommended. In the event vehicles are not used to return to the visitor's center, the quickest route is to follow the trace of the New Garden Road due west.

---

[103] "Follow on Report from General Cornwallis to Lord George Germain on the Battle of Guilford Courthouse, 17 March 1781," in Goss, *Primary Source Reference Book for the Guilford Courthouse Campaign*, 11–14.
[104] *Offense and Defense*, 1-7–1-9; and *Marine Corps Operations*, 6-1–6-2.

PART IV

INTEGRATION

This chapter provides a suggested framework for conducting the all-important integration phase of the staff ride. As outlined in the introduction, integration should occur as soon as possible to allow students to capture, synthesize, and articulate their observations and insights. To omit or rush this portion is to miss the entire point of the staff ride: What did I learn, and how do I apply what I learned today to improve myself and/or my profession?[1] Before the start of the field phase event, the facilitator should ensure the training audience clearly understands the staff ride is a training event and not just an interesting terrain walk. Here, the facilitator should make every effort to have the unit commander clearly articulate training objectives and goals for the event. During the terrain walk, periodically remind students that there will be a test at the end as a means of motivating active participation in the sharing of insights as they form. The facilitator should encourage note-taking as a way of capturing fleeting thoughts and impressions for later sharing. One helpful technique is to appoint a recorder to take notes (or even record the event with a digital voice and/or video recorder) of the discussion points from each stand and review the notes at the end.

In planning and preparing for the integration phase, the facilitator must balance some competing factors. Ideally, the integration session is held at a location with minimal environmental and noise distractions. Sufficient time to conduct the integration is necessary, so the facilitator should coordinate in advance with the unit commander to ensure an adequate amount of time is blocked out on the training schedule. When possible, students should have time for personal reflection and thought before the integration phase. Therefore, the optimum timing is to hold the integration session the day following the field study phase.

However, few units will afford the staff ride leader the luxury of extra time, so the leader must prepare to conduct the integration phase immediately following the field study. Several considerations are in order. After a couple of hours walking the battlefield, the physical requirements of all participants (e.g., food, water, and restroom relief) will certainly impact the ability to perform critical thinking. Also, the staff ride facilitator should organize the integration phase based on the type of unit—the focus of a brigade will likely differ from that of a division- or corps-level formation—time available, and training objectives and goals of the unit commander. Address the commander's expectations for the integration event beforehand, as a formal event may require the coordination of the auditorium space and audiovisual support. Conversely, an informal integration event can easily take place in conjunction with a post–staff ride meal. At minimum, the integration event should take place at a location different from the last stand and in a setting conducive to open discussion among all participants. The staff ride leader can employ a structured or unstructured format, but whatever method is used, the staff ride leader should facilitate the event and let the students do the majority of the talking. By doing so, the students ideally integrate their preliminary study with insights gained on the terrain walk to leverage relevant insights into their current assignment as well as subsequent career. At this juncture, it is important to understand that the integration phase does not include an after action review or report (AAR) of the staff ride. Although important, AAR comments fall outside the scope of the integration event and should be done separately.

To assist the facilitator in the process, this section provides some possible techniques along with

---

[1] Maj Matthew Cavanaugh, "The Historical Staff Ride, Version 2.0: Educational and Strategic Frameworks" (professional development paper, U.S. Military Academy, 2014), 4–6. Cavanaugh provides excellent pointers on how to ensure the staff ride is conducted with sufficient historical rigor.

sample questions that can be used to cap a quality training event. Figures 2, 3, and 4 (pp. 126–127) also provide additional analysis points using Joint- or Service-specific doctrine as a structuring framework. The questions should help students link their experiences and observations to the commander's stated training objectives. At the beginning of the integration event, the facilitator should remind students of the goals and objectives of the training event and the purpose of the integration phase. Then the facilitator should pose open-ended questions to help stimulate a robust question and comment session. Depending on the audience and motivation level, the facilitator may need to do little more than ask the question and let the conversation run its course. In some instances, the facilitator may have to ask focused follow-up questions to draw out insights from more reserved participants. One good method is to ask the following open-ended questions:

1. What initial impressions of the campaign did you develop in the preliminary study phase that were either changed or strengthened after the terrain walk?

This question gets to the heart of a staff ride: the study of the terrain in relation to the course of the battle. Some facets of the discussion could include, for example, the vast distances of the operational maneuver leading to the battle and the impact of the rugged terrain on the operational and tactical maneuvers. Good follow-up questions here are: Did seeing the terrain alter your opinion of any of the leaders? If so, how?

2. What aspects of warfare have changed since the Battle of Guilford Courthouse? What have remained the same?

The obvious changed aspects will generally revolve around technology: weapons, mechanization and motorization, communications, and support equipment. More subtle, but discoverable with a bit of encouragement from the facilitator, are the timeless aspects of warfare: the role of personalities, relationships, commander's intent and guidance, logistics, and the presence or absence of assertive leadership on the battlefield.

3. What relevant insights can a military professional glean from the study of the Guilford Courthouse battle?

This question can easily open myriad discussion threads, limited only by time and student interest. Here, the facilitator can help frame the discussion based on the type of unit. For example, a logistics unit might find much insight in comparing and contrasting how Generals Greene and Cornwallis organized their logistics systems. Although this staff ride incorporates much in the way of tactical-level analysis, it lends itself well to operational-level considerations. Consequently, a potential operational-level discussion framework is found in the elements of operational design:

    a. end state/military conditions
    b. center of gravity
    c. decisive points and objectives
    d. lines of operation
    e. culmination point
    f. operational reach/approach/pause
    g. simultaneous and sequential operations
    h. tempo

For tactical-level analysis, the Army's six and the Marine Corps' seven warfighting functions (WFF) can serve as a good launching point for questions about:

    a. maneuver

    b. intelligence

    c. fires

    d. logistics

    e. force protection

    f. command and control

    g. information

These questions and focus areas are provided simply as an aid to stimulating a robust integration phase, not as a prescriptive list. This handbook provides examples of possible answers to the questions but does not purport to provide the right answer to operational and tactical problems. In preparation for the event, the staff ride leader should prerecord answers or thoughts to the questions to have some potential ideas to kick-start the discussion, if necessary. Ideally, the students will engage in a spirited discussion that will require minimal instructor input.[2]

---

[2] The author also recommends consulting Peter G. Knight and William G. Robertson, *The Staff Ride: Fundamentals, Experiences, Techniques* (Washington, DC: U.S. Army Center of Military History, 2020). The publication has been extensively updated with detailed planning, guidance, and staff ride pedagogy; it was unfortunately released too late to incorporate into this work.

AREA OF ANALYSIS	GREAT BRITAIN	AREA OF ANALYSIS	UNITED STATES
Strategic Goals	Regain control of North America Protect Caribbean holdings	**Strategic Goals**	Exhaust the British will Gain foreign recognition and intervention Gain independence
Operational Goals	Destroy/disperse the Continental Army Pacify the Carolinas *Protect Loyalists?*	**Operational Goals**	Prevent British consolidation Pacify the Carolinas Establish American sovereignty
Tactical Objectives	Occupy key locations Organize Loyalist militias Destroy/neutralize rebels	**Tactical Objectives**	Destroy isolated British and Loyalist units Disrupt recruiting of Loyalists Deny provisions to the enemy
Center of Gravity	Support of local population Effective Loyalist militia	**Center of Gravity**	Support of local population Viable Continental presence in the Carolinas
Lines of Operation	Exterior seaborne lines of operation Exterior land lines of operation	**Lines of Operation**	Interior lines of operation No use of sea lines of operation due to Royal Navy presence
Decisive Points	Discipline Volley fire and the bayonet charge	**Decisive Points**	Leadership Motivation Marksmanship
Culminating Points	Loss of Loyalist support Loss of offensive capability in regular units	**Culminating Points**	Loss of Patriot support Destruction of Continental Army in the Carolinas

GUILFORD COURTHOUSE
BRITISH ARMY

AREA OF ANALYSIS	CORNWALLIS'S ARMY	AREA OF ANALYSIS	CORNWALLIS'S ARMY
Movement and Maneuver	Movement in massed columns Deployment of infantry into line formation Bayonet Charge to decide the battle	Force Protection	Battle drills Use of mobile troops and pickets to protect bivouac sites No use of light infantry as skirmishers to prevent surprises during move to contact
Intelligence	Lack of terrain knowledge hinders pursuit Intelligence from locals inaccurate or deliberately misleading Cornwallis performs no leader's recon before attacking at Guilford Courthouse	Command and Control	Cornwallis issues no branch or sequel plans to guide his commanders No synchronization of regimental actions after first contact Cornwallis and his subordinate officers lead from the front
Fires	Superb fire discipline Massed volley fires for shock effect Employment of jaëgers to attrite leadership	Information	Cornwallis's letters to Clinton are vague about his intent and actions
Logistics	Regular supply lines nonexistent Destruction of logistics trains hinders ability to carry supplies Austere ordnance and hospital trains	Civil Considerations	Little effort to protect Loyalist militia and populace from Patriot retribution Cornwallis's army does not remain in one location to provide security

GUILFORD COURTHOUSE
CONTINENTAL ARMY

AREA OF ANALYSIS	GREENE'S ARMY	AREA OF ANALYSIS	GREENE'S ARMY
Movement and Maneuver	Greene's forces move dispersed to reduce vulnerability Continental movement and maneuver similar to that of the British Army	Force Protection	Corps of observation, pickets used to screen movements and encampments Open order tactics, use of natural cover and concealment No prepared obstacles or fortifications
Intelligence	Unmatched knowledge of area of operations from engineer and personal reconnaissance Effective use of local spies and guides	Command and Control	Greene does not appoint an overall commander for first and second line defenses Greene and his subordinate Continental officers lead by example at critical points during the battle
Fires	Riflemen target officers and NCOs to disrupt command and control 6–pounders used in support of first line defense	Information	Greene writes detailed reports to General Washington and key political leaders Makes operational level decisions influenced by information considerations
Logistics	Resupply via interior lines from Virginia Militia sweep up provisions and forage to deny resupply to British	Civil Considerations	Disciplines his soldiers to control looting Discourages ill treatment of Loyalists Orders the issuance of receipts for all requisitioned supplies

PART V

1. Information and assistance

a. The Staff Ride Team, Combat Studies Institute (CSI), of the Army University Press (AUP) at Fort Leavenworth, Kansas, can provide assistance and advice on the planning and execution of a professional staff ride. Visit the CSI website to obtain additional information on staff ride assistance:

>   Address: Army University Press/Combat Studies Institute
>   ATTN: ATZL-CSH
>   201 Sedgwick Avenue, Building 315
>   Fort Leavenworth, KS 66027
>   Telephone: DSN: 552-2078
>   Commercial: (913) 684-2131
>   Website: http://www.armyupress.army.mil/Educational-Services/Staff-Ride-Team-Offerings/

b. For Army Reserve units, the Office of Army Reserve History can assist with planning and executing staff rides:

>   Address: Army Reserve Headquarters
>   ATTN: Office of Army Reserve History
>   4700 Knox Street
>   Fort Bragg, NC 28310
>   Commercial: (910) 570-8194
>   Website: https://www.usar.army.mil/ArmyReserveHistory/

c. The U.S. Army Center of Military History develops and leads staff rides for U.S. Army groups and can help support other official government agencies and departments.

>   Address: United States Army Center of Military History
>   ATTN: Chief,
>   Field and International History Programs Division
>   102 4th Avenue, Suite 100
>   Fort Lesley J. McNair, Washington, DC 20319-5060
>   Commercial: (202) 685-2726, (202) 445-0570
>   Organizational email: usarmy.mcnair.cmh.mbx.answers@mail.mil
>   Website: https://history.army.mil/staffRides/index.html

d. The Guilford Courthouse National Military Park (GCNMP) is located in Greensboro, North Carolina, in close proximity to Interstate 40 and the Piedmont Triad International Airport. Unlike many other American battlefields, the Guilford Courthouse battlefield sits entirely within the city limits of Greensboro, so all desired amenities for before or after a staff ride (e.g., gas, lodging, and restaurants) are easily found in close proximity to the park. GCNMP is under the administrative control of the National Park Service (NPS) and features a full-service visitor's center with museum displays, bookstore, a 110-

seat theater, and a 30-seat education center. The museum rates as one of the best in the NPS inventory, with an extensive collection of artifacts and interpretive displays. If time permits, the 30-minute video presentation *Another Such Battle*, the official NPS history of the battle, can serve as an excellent introduction to the topic. Besides the video, the center features an animated battle map presentation that runs approximately 10 minutes. The park visitor's center is open daily with the exception of Thanksgiving, Christmas, and New Year's Day, from 0830 to 1700, and admission is free. The park is accessible from sunrise to sunset. For more information, contact the park staff at:

      Address: Guilford Courthouse National Military Park

             2332 New Garden Road

             Greensboro, NC 27410

      Telephone: (336) 288-1776

      Website: www.nps.gov/guco

e. The Cowpens National Battlefield is located near Gaffney, South Carolina. Unlike Guilford Courthouse, the battlefield is in an isolated, rural area off Interstate 85 (I-85), and services are located about 10 miles (16 km) away at the intersection of SC-11 and I-85 at Gaffney, South Carolina. The park is administered by the NPS, with a full-service visitor's center, bookstore, a roomy theater suited for classes, and a modest museum. A 30-minute video presentation will give a good tactical-level view of the Cowpens battle, but it includes only limited strategic and operational information. The park visitor's center is open daily, except on major federal holidays, from 0900 to 1700 year-round, and admission is free. For more information contact the park superintendent at:

      Address: Cowpens National Battlefield

             4001 Chesnee Highway

             Gaffney, SC 29341

      Telephone: (864) 461-2828

      Website: www.nps.gov/cowp

2. Logistics for the Guilford Courthouse National Military Park

a. Meals and services. The battlefield is located less than 1 mile (1.6 km) from a major shopping area with several restaurants, convenience stores, and gas stations.

b. Lodging. The closest available commercial lodging is available at the nearby Piedmont Triad International Airport, Greensboro, North Carolina, around 7 miles (11 km) from the park.

c. Traveling. Access to the park is gained by taking Interstate 40 to Interstate 73 north to the Joseph M. Bryan interchange. From there, travel east to New Garden Road, and the entrance to the park is approximately 2.5 miles (4 km) from that point. Piedmont Triad International Airport is the closest regional airport for groups flying from an extended distance.

3. Suggested logistics arrangements for the campaign-level stands

a. McGuire Nuclear Power Plant Viewing Point. There are no restrooms or potable water sources at the viewing point. Services are available to the east at the intersection of NC-

73 and Interstate 77, about 5 miles (8 km) to the east, or about 3 miles (4.8 km) to the west at the intersection of NC-73 and NC-16.

b. Yadkin River Crossing Point. There are no restrooms or potable water sources at the river location. Services are available in East Spencer, North Carolina, just south of the site on US-52/I-85.

c. Boyd's Ferry. No services at the riverside, but services are available within the community of South Boston, Virginia.

d. Whitesell's Mill. There are no restrooms or potable water sources at the location. The closest services are found adjacent to I-85 at Burlington, North Carolina, about 10 miles (16 km) southeast on NC-61.

e. New Garden Meeting House. This location is in close proximity to the Guilford Courthouse battlefield, so simply follow the details outlined in section 2 above.

4. Other considerations

a. As with any other training event, personal reconnaissance of the stands is essential to a successful event. With advanced notice, NPS staff at both Cowpens and Guilford Courthouse have the ability to support a staff ride with assistant instructors and a firing demonstration.

b. The main visitor's center at Guilford Courthouse has only two water fountains and a single water bottle refill station, while the modern toilet facilities are designed for a limited number of visitors; consequently, unit planners should account for the limited water and toilet facilities in their logistics plans. A seasonal toilet facility with a single water refill station is located on the east side of the battlefield at the parking lot for the third line.

c. Parking at the Guilford Courthouse visitor's center is adequate, with three spots for tour bus-size vehicles. However, the park and adjacent areas serve as a popular greenspace for local residents. During periods of good weather, parking will fill up quickly, and the main trails to the east of the visitor's center might be crowded during peak hours. The parking lot adjacent to the third line has 18 automobile-size spaces that are time limited, so this parking lot is best suited as a post-event pickup point.

d. The ideal time for a visit to Guilford Courthouse is in March so as to best replicate the weather, light, and vegetation conditions of the battle. Otherwise, the recommended time to visit is during the milder late fall to early spring periods, when the trees are not full of leaves, which would impede clear line of sight on the battlefield.

e. All members should wear hiking boots or shoes and long pants and carry rain gear and insect repellent. Many of the stands will take place in or adjacent to wooded and/or swampy areas, so precautions for poison ivy, mosquitos, and other biting insects are necessary.

APPENDIX A

Assigning selected students to each study one of the key participants of the Guilford Courthouse battle can accomplish several goals for the staff ride. First, assigning a read-ahead activity should ensure that students research the characters enough to avoid embarrassment on the day of the event. Ideally, each student will perform an in-depth analysis of the historical character to be able to discuss why the character acted in a particular way and how the actions impacted the outcome of the battle.

Second, role-playing nests well within the adult learning pedagogical model; students learn best when they form a personal interest or attachment to the subject. Study of the historical participants can create an intellectual and emotional connection and stimulate a student's interest in the past and ideally foster a career-long desire for additional self-study and reflection on the profession of arms.

Last, role-playing can inject some levity into an otherwise serious and often grim subject, thus serving to lighten the mood and maintain participant interest and motivation.

All ranks in biographies are given as the day of the Guilford Courthouse battle on 15 March 1781.

AMERICANS
Colonel William Campbell

Born in 1745, Colonel William Campbell was one of the leading citizens of Washington County, Virginia. Before the Revolution, Campbell had gained extensive experience leading militia units against indigenous tribes. After the outbreak of war in 1775, Campbell was commissioned a captain in the 1st Virginia Regiment and served briefly in the Continental Army. After resigning his Continental commission, Campbell married Elizabeth Henry, a sister of Patrick Henry, and was appointed as Washington County's militia colonel. During the middle stages of the Revolutionary War, Campbell successfully commanded expeditions against both Tory and native threats, all while serving as a member of the House of Delegates. During the campaign to destroy Major Patrick Ferguson's *Loyalist Corps of Militia*, Campbell led a regiment of 400 Virginia riflemen who fought with distinction at Kings Mountain in October 1780. After returning home in November 1780, Campbell commanded Virginia militia units performing local defense against frequent Loyalist and Cherokee raids. The need to maintain a sizable militia presence in the state meant that Campbell could only lead an understrength battalion of riflemen to support Major General Nathanael Greene at Guilford Courthouse. Campbell was promoted to brigadier general in 1781 and given command of a Virginia militia brigade that reinforced the Continental Army during the Yorktown campaign. There, Campbell sickened, possibly from malaria, and died on 22 August 1781 at the age of 36. In 1822, well after the war when Campbell lay moldering in the ground, John Sevier and Isaac Shelby publicly questioned Campbell's performance on the battlefield in a series of letters and pamphlets, accusing him of shirking during the height of the battle. Relatives and members of Campbell's command wrote letters defending his actions at Kings Mountain, while providing evidence to support a case of mistaken identity.[1]

---

[1] Lyman C. Draper, *King's Mountain and Its Heroes: History of the Battle of King's Mountain, October 7th, 1780, and the Events Which Led to It* (Cincinnati, OH: Peter G. Thomson, 1881), 378–402.

Major General Nathanael Greene

An unlikely warrior and battlefield commander, Nathanael Greene earned a reputation as second only to George Washington as a hero of the American Revolution. Greene was born in 1742 to a hardworking and prosperous Quaker family in Potowomut, Rhode Island. At an early age, Greene displayed a remarkable intellect despite a very limited formal education. During his youth, Greene studied Latin and mathematics and read regularly between stints working at a forge and gristmill. He amassed a considerable library, and in his early adult years studied books on military matters, particularly the writings of Henri de La Tour d'Auvergne, vicomte de Turenne and marshal of France; Frederick the Great; and Maurice, comte de Saxe and marshal general of France. In February 1772, the HMS *Gaspee*, a British customs schooner captained by Lieutenant William Dudingston, boarded a Greene family trading sloop and confiscated 12 hogsheads of undeclared rum. The incident angered most Rhode Islanders, including Nathanael Greene, and in 1774 Greene helped to organize an independent militia company in East Greenwich, Rhode Island. Despite his role in setting up the company, Greene was denied an officer's commission due to physical limitations (a bad knee and chronic asthma). Although privately embarrassed at the rejection, Greene remained in the company as a private soldier. Remarkably, when the Rhode Island militia was called out for the siege of Boston, Greene was selected from among the ranks as a brigadier general. General George Washington soon noticed Greene's innate intelligence and grasp of military strategy and soon appointed the 32-year-old Greene as a brigadier general of the Continental Army, making him the youngest general officer in the new American force. Greene soon found the reality of war much different than that presented in books. He fared well in his first battlefield command at Harlem Heights in September 1776 but erred badly by insisting his troops could hold Fort Washington, guarding the north end of Manhattan Island. A British and Hessian army easily took the fort on 16 November 1776, capturing 2,838 Continentals, 34 cannons, and several tons of ordnance stores. Despite Greene's major role in the disaster, Washington did not lose confidence in Greene, a trust that was repaid during the Continental Army's subsequent withdrawal across New Jersey, where Greene's logistics arrangements kept the Army reasonably well fed and supplied. During the winter of 1776–77, Greene showed he had the mental hardness for waging war, as he successfully cleared the countryside of all animals necessary to supply meat and draft animals to the Continental Army. To Washington, he reported, "The inhabitants cry out and beset me from all quarters, but like Pharoh I harden my heart."[2] In February 1778, Greene was appointed quartermaster general of the Continental Army, a position he would hold until taking command of the Southern Department in December 1780. Greene resented the appointment, noting ruefully to General Washington, "No body ever heard of a quarter Master in History." Yet, Greene made significant improvements in the Continental logistics systems, establishing a system of field depots and primitive factories to supply uniforms, ordnance stores, and equipment to field armies. Greene also improved transportation methods and used engineer officers to survey and improve transportation routes. During his tenure as quartermaster general, Greene also managed to squeeze in some combat time, successfully commanding Continental troops at the battles in Monmouth and Rhode Island during 1779. After a dustup with Congress about a reorganization of the Quartermaster Department, Greene resigned as quartermaster general but was soon given command of the Southern Department by George Washington. Greene commanded the Southern Department from December 1780 to the end of the war. Despite losing almost every battle it fought, Greene's army ultimately regained control of the Carolinas and Georgia, except the port cities of Charleston and Savannah, by the end of 1781. The postwar years were not kind to Greene, as he

---

[2] MajGen Nathanael Greene to Gen George Washington, 15 February 1778, Founders Online pages, National Archives and Records Administration, accessed 8 April 2020.

was burdened by debts incurred during his time as the quartermaster general, and his efforts to turn a profit from a rice plantation ended prematurely on 19 June 1786, with his sudden death from complications caused by heatstroke.[3]

Lieutenant Colonel John Eager Howard

John Eager Howard was born into a Maryland planter's family in June 1752. During his youth, he completed a classical education. In July 1776, Howard declined a colonelcy due to his lack of military experience, instead accepting a captain's commission in the 2d Maryland Brigade of the Flying Camp, a sort of mobile reserve unit. On 10 December 1776, Howard was commissioned major of the 4th Maryland Regiment, and later promoted as lieutenant colonel of the 5th Maryland Regiment on 11 March 1778. During his time in the Northern Department, Howard participated in battles at White Plains, Germantown, and Monmouth. In 1780, Howard marched south with the 2d Maryland Regiment as reinforcements sent to the Southern Department. After the Camden disaster, Howard took command of the combined Maryland-Delaware regiment and led it with distinction at the Cowpens in January 1781, for which he received a silver medal from Congress. Howard led Continental forces at the remaining battles of the Southern Campaign, during which he was severely wounded at Eutaw Springs, South Carolina, in September 1781. Despite his debilitating wartime wounds, Howard served in the Continental Congress and later as governor of Maryland and as a senator. Howard died in Baltimore, Maryland, on 12 October 1827.[4]

Brigadier General Isaac Huger

Isaac Huger (pronounced YOU-gee) was born in March 1742 to a wealthy Huguenot merchant and plantation owner on the Cooper River of South Carolina. In his youth, Huger was sent abroad to Europe for a formal education. After his return to South Carolina, Huger joined the Provincial South Carolina Regiment as a company grade officer and fought his first combat at age 18 against a band of Cherokee. At the start of the Revolution, Huger was commissioned lieutenant colonel of the 1st South Carolina Regiment and in September 1776 was commissioned colonel of the 5th Continental Regiment. By 1779, Huger was a brigadier general, leading Continental troops in the unsuccessful defense of Georgia against a British offensive from east Florida. During the Stono Ferry battle in June 1779, Huger suffered his first combat wound. Huger was in charge of a brigade of light horse and Carolina militia during the siege of Charleston, in which his command was beaten and dispersed by Tarleton's *Legion* at Monck's Corner, South Carolina, in April 1780. Huger was subsequently evacuated due to sickness, thus avoiding capture when Charleston surrendered in May 1780. After his recovery, Huger took command of a new brigade of Virginia Continental troops, marching them south to reinforce the Southern Department in the fall of 1780. Huger led the Virginia Continental brigade during the Guilford Courthouse fight in March 1781, where he was wounded in the final stages of that battle. After the war, Huger was elected to the South Carolina General Assembly, and in September 1789 he was appointed the first U.S. Marshal of South Carolina. Poor health and personal matters led to Huger's retirement in 1793, and he died in October 1797.[5]

Lieutenant Colonel Henry "Light-Horse Harry" Lee III

Born to a Virginia family near Dumfries, Virginia, Henry Lee studied law at the College of New Jer-

---

[3] William Johnson, *Sketches of the Life and Correspondence of Nathanael Greene, Major General of the Armies of the United States in the War of the Revolution*, vol. 1 (Charleston, SC: A. E. Miller, 1822).
[4] John Eager Howard (1752–1827), MSA SC 3520-692, Governor of Maryland, 1788–1791, Biographical Series, Archives of Maryland, accessed 8 April 2020.
[5] "Isaac Huger, Quick Facts," National Park Service, accessed 8 April 2020.

sey in Ewing, graduating in 1773. A naturally skilled and daring equestrian, Lee was commissioned captain of a troop of Virginia light dragoons in 1776, which was attached to the 1st Continental Light Dragoons for service during the Philadelphia Campaign. Lee's aptitude led to a promotion to major and command of a newly organized mixed regiment of infantry and cavalry called Lee's Partisan Legion. Lee proved a good battlefield commander, defeating a Hessian regiment at Edgar's Lane and capturing a British battalion with the loss of only one fighter during the Battle of Paulus Hook, New Jersey, in August 1779. Lee's exploits in the Northern Department earned him the epithet "Light-Horse Harry," promotion to lieutenant colonel, and the award of a gold medal from Congress, the only such medal bestowed on a field grade officer during the war. Lee's Partisan Legion was sent as a reinforcement to the Southern Department after the Camden disaster, where it served with distinction during the remainder of the war. After the war, Lee served in the Virginia House of Delegates and was governor of Virginia for three terms. Debts accrued by bad land deals landed Lee in debtor's prison in 1807, where he wrote his *Memoirs of the War in the Southern Department of the United States*. In 1812, Lee was caught up in a mob action over the War of 1812 and received head and body trauma that affected his health and speech. Lee visited the West Indies hoping to recover his health in the tropical warmth, but his health declined, and he died at Dungeness (at the home of Mrs. Shaw, the daughter of his former commander, Greene) on Cumberland Island, Georgia, on 25 March 1818, en route to his home in Virginia.[6]

Brigadier General Edward Stevens

Edward Stevens was born in Culpeper County, Virginia, in 1745. Stevens began his military service as lieutenant colonel of the Culpeper battalion of minutemen raised in 1775, a detachment of which he led at the Battle of Great Bridge on 9 December 1775. In November 1776, Stevens was commissioned colonel of the 10th Virginia Regiment of Continentals. Stevens led his regiment with distinction during the Philadelphia Campaign before resigning his commission in January 1778. Stevens recruited a battalion of the Virginia Line in June 1778 and was subsequently appointed brigadier general of the Virginia militia in 1779. He commanded Virginia Line brigades at Camden in August 1780 and at Guilford Courthouse in March 1781, where he was severely wounded. After a period of convalescence, Stevens returned to action, leading a Virginia brigade during the Virginia campaigns in 1781. Stevens served brief terms as a state senator in 1776 and 1779–80 before dying in 1820.[7]

Major St. George Tucker

Probably one of the most colorful characters to write about his experiences during the Revolution, St. George Tucker was born at Port Royal, Bermuda, in 1752. Raised in a wealthy trader family, Tucker remained in Bermuda until age 19 before attending the College of William and Mary in Williamsburg, Virginia, to study law. Shortly after Tucker passed his bar exam, the outbreak of the revolution led to the closure of the Virginia courts, and he was forced to return to Bermuda. There, Tucker became involved in smuggling to help the Continental cause. One of his largest projects was the covert seizure of 100 barrels of gunpowder from the royal powder magazine at Bermuda, which was successfully smuggled off the island to resupply the Continental Army. Tucker returned to Virginia in January 1777, landing at Yorktown in a family-owned cargo ship carrying smuggled salt used to preserve meat for the Continental Army. Tucker remained in Williamsburg, where he arranged for the shipment of American commodities to the West Indies in exchange for firearms and ordnance stores.

6 Henry Lee, *Memoirs of the War in the Southern Department of the United States* (New York: University Publishing, 1869), 16–56.
7 Thomas R. Jones, *A New American Biographical Dictionary, or Remembrancer of the Departed Heroes, Sages and Statesmen of America* (Philadelphia, PA: Samuel F. Bradford, 1829), 373–76.

In 1779, Tucker was commissioned major in Colonel John Holcomb's Virginia militia regiment, comprising companies from Amelia, Charlotte, Mecklenburg, and Powhatan Counties. Tucker's first exposure to combat was probably at Guilford Courthouse, as no record exists of any field service before the battle. During the wreck of Holcombe's regiment, Tucker rallied a company's worth of troops who "sustained an irregular kind of skirmishing with the British."[8] At the close of the battle, Tucker was a victim of friendly fire when, as he wrote, he was "attempting to rally a party of regular Troops . . . received a wound in the small of my Leg from a soldier, who either by design or accident held his Bayonet in such a direction that I could not possibly avoid it as I rode up to stop him from running away."[9] Tucker recovered after the battle and went on to command a Virginia regiment that fought at Yorktown, where he was wounded a second time by British shellfire. After the war, Tucker resumed his law practice in Petersburg, Virginia, and was later appointed as a professor of law at his alma mater and as a judge at the Virginia General Court. Among his other varied accomplishments, Tucker edited William Blackstone's *Commentaries on the Laws of England* (1765) and published a pamphlet advocating the gradual abolishment of slavery. Tucker died in November 1827.[10]

Colonel William Washington

A distant relation to George Washington, William Washington was born in Stafford County, Virginia, in February 1752. At the outbreak of the war, Washington abandoned his theological studies to obtain a captain's commission in the 3d Virginia Continentals on 25 February 1776. Washington served with honor during the New Jersey and New York campaigns, suffering wounds at the Battles of Long Island and Trenton. After Trenton in December 1776, Washington transferred to the dragoons and was soon promoted to major along with command of the 4th Continental Dragoons. On 20 November 1778, Washington was promoted to lieutenant colonel and given command of the 3d Continental Dragoons. With the shift of Continental reinforcements to South Carolina, he led his squadron south to Charleston. During skirmishes around Charleston, Washington's squadron was badly handled by Tarleton's *Legion*. First, Washington lost most of his mounts and equipment at Monck's Corner. The second time, Washington's reconstituted command was driven into the Santee River, with Washington and his officers having to swim the river to safety. After that, Washington withdrew his squadron to North Carolina to replace his losses. In December 1780, the 3d Dragoons were amalgamated with South Carolina militia dragoons in Morgan's flying army. Washington's performance was much improved and his dragoons won victories at Rugeley's Mill and Hammond's Store. At the Battle of Cowpens, Washington ably commanded a squadron of 130 dragoons. Numerically outnumbered by the dragoons of Tarleton's *Legion*, Washington kept his forces in a compact mass and decisively completed Morgan's envelopment of the British infantry to end the battle. Washington was awarded a silver medal by Congress in recognition of his accomplishments at Cowpens. During the Race for the Dan, Washington's dragoon squadron helped screen the main body of Continentals, wrecking bridges and fighting delaying actions against the British van. At Guilford Courthouse, he led a corps of observation that screened the American right flank and fought a delaying action back to the Continental third line that delayed and disrupted the British advance on the main Continental line. At the height of the battle, Washington led a well-timed attack that disrupted the British *2d Guards* battalion, giving General Greene sufficient time to withdraw his Continental regiments. After Guilford

---

[8] Charles Washington Coleman Jr., ed., "The Southern Campaign, 1781, from Guilford Court House to the Siege of York, Narrated in the Letters from Judge St. George Tucker to His Wife, the Southern Campaign as Narrated by St. George Tucker," in *The Magazine of American History*, vol. 7 (New York: A. S. Barnes, 1881), 40.
[9] Coleman, "The Southern Campaign, 1781, from Guilford Court House to the Siege of York, Narrated in the Letters from Judge St. George Tucker to His Wife, the Southern Campaign as Narrated by St. George Tucker," 39–40.
[10] Davison M. Douglas, "St. George Tucker (1752–1827)," Encyclopedia Virginia (website), accessed 8 April 2020.

Courthouse, Washington ran into trouble at the Battle of Eutaw Springs, where he was bayonetted and captured on 8 September 1781. Despite his occasional battlefield reverses, usually brought on by lax security measures, Washington was esteemed as a gifted battlefield commander. Cornwallis later noted that "there could be no more formidable antagonist in a charge, at the head of his cavalry, than Colonel William Washington." After the war, Washington settled on the Sandy Hill plantation near Charleston, South Carolina. He served a couple of sessions in the state assembly and was appointed a brigadier general in the state militia. Washington died in Charleston in March 1810.[11]

Lieutenant Colonel Otho Holland Williams

Otho Holland Williams was born in Prince George's County, Maryland, in 1749 in a farming family of Welsh descent. At 13, Williams began an apprenticeship in the Fredrick County clerk's office, where he eventually rose to the position of lead clerk. At 18, he moved to Baltimore where he continued the clerking profession. In 1775, Williams obtained a lieutenant's commission in an independent rifle company that served at the siege of Boston. When the company was incorporated into the Continental Army in 1776, Williams was promoted as major of the Maryland and Virginia Rifle Regiment, a single rifle regiment formed by combining Virginia and Maryland rifle companies into a composite unit. He was captured at Fort Washington and imprisoned for a time in New York, where lack of food and mistreatment permanently affected his health. Williams was exchanged in January 1778 following the American victory at Saratoga and was immediately commissioned as the colonel of the 6th Maryland Regiment. In April 1780, the 6th Maryland Regiment was sent south as part of reinforcements under the command of General Johann de Kalb. Besides commanding his own regiment, Williams put his civilian clerking skills to good use, serving as de Kalb's deputy adjutant general. During the opening portion of the Battle of Camden, Williams led a group of riflemen skirmishers that unsuccessfully tried to slow down the British deployment for battle. Even after the American militia on the left wing gave way in a panic, Williams continued to shuttle orders between de Kalb and the subordinate Continental brigades and regiments during their final desperate efforts to avoid disaster. After the battle, Cornwallis claimed to have inflicted 1,000 casualties and captured about 800 Continentals. Williams was one of the lucky ones, escaping the wreck of the American Army by fleeing deep into the pine barren swamps. Williams soon rejoined the much-depleted Southern Army at Hillsborough, North Carolina, where he took command of the amalgamated Delaware-Maryland Continental brigade in addition to his day-to-day adjutant duties. When General Daniel Morgan's flying army was dispatched into South Carolina, a battalion of light infantry went forward, and the remainder of the Delaware-Maryland brigade remained behind under Williams's command. When Greene's army reunited after Cowpens, Williams took command of the consolidated light infantry corps, which he skillfully led as the rear guard of the American Army during the Race for the Dan; he was later relieved from this command when Greene reorganized the light troops after the Whitesell's Mill skirmish. At Guilford Courthouse, Williams led the Delaware-Maryland brigade in battle. After directing the successful attack of the 1st Maryland into the *Guards*, Williams was caught up in the collapse of the 2d Maryland, and he narrowly avoided death or capture when he unsuccessfully tried to regroup his panicked troops into formation. Once General Greene ordered the Continental forces to withdraw, Williams successfully extricated his brigade without further personnel losses. Williams remained dual-hatted as the adjutant general and commander of the Maryland brigade to the end of the war. After the war, Williams was a major presence in the city of Baltimore, marrying into one of the city's wealthiest merchant families and for a time managing the Port of Baltimore. Williams's health in later years declined on account of his imprisonment and arduous service in the Southern Campaign, and he died in July 1794.[12]

BRITISH
Lieutenant General Lord Charles Cornwallis

The central British figure in the Southern Campaign, Lieutenant General Charles Cornwallis, first marquess and second earl Cornwallis, was among the best battlefield commanders fielded by the British Army during the war. Cornwallis was born into the aristocracy in 1738, attending Eton College in Windsor and Cambridge University in his youth. In 1756, his father purchased an ensign's commission in the elite *1st Foot Guards* for him. In 1757, Cornwallis took a leave of absence to tour the battlefields of Europe with a military tutor and study military science at the Turin military academy in Italy. At the outbreak of the Seven Years' War, Cornwallis missed deploying to Europe with the *1st Guards*; instead, he wrangled a staff officer appointment with the British commander in chief. Cornwallis first saw combat at Minden in 1759 and soon after purchased a captaincy in the *85th Regiment of Foot*. In 1761, he was commissioned in the *12th Foot* and subsequently brevetted to lieutenant colonel for his excellent combat leadership. While he was still leading troops in combat, Cornwallis was selected for the House of Commons in 1760 and elevated to the House of Lords on the death of his father in 1762. Favorable court connections led to his appointment as the colonel of the *33d Regiment of Foot* in 1766, but busy with political matters, Cornwallis saw no combat service for the intervening decade. In 1776, Cornwallis was promoted to major general and assigned as a corps commander in General William Howe's army during the New York campaign. Cornwallis performed superbly during the campaign, badly battering the inept Continentals and driving them from New York, but he badly fumbled the pursuit of Washington's army after its surprising victory at Trenton, New Jersey. Cornwallis shrugged off a sharp rebuke from General Howe and performed well in the capture of Philadelphia and in the battles of Brandywine, Germantown, and Fort Mercer. However, Cornwallis's seemingly half-hearted engagement of the enemy at Monmouth in June 1778 brought some criticism from Lieutenant General Sir Henry Clinton. Rancor festered between the two ever afterward and particularly affected their relationship during the latter stages of the war. During the siege of Charleston, Cornwallis led a corps of infantry that helped close the cordon around the city. Afterward, Clinton returned to New York as commander in chief, leaving Cornwallis in independent command of all British forces in the South. With Clinton gone, Cornwallis felt free to run the war the way he saw fit. At first, Cornwallis seemed very successful as his regular troops won several victories over Whig insurgents and with the major victory at Camden in August 1780. He pursued an aggressive strategy against the American guerrillas, even employing his newly formed Tory militia in an offensive capacity—an experiment that ended in disaster at Kings Mountain in October 1780. Bereft of his rear and flank security, Cornwallis was forced to suspend an invasion of North Carolina to handle the growing American threat. He split his army in an attempt to trap and destroy the Continental light forces under Colonel Daniel Morgan. Instead, trapped by Morgan at the Cowpens in January 1780, Cornwallis lost most of his light infantry. Fixated on running Major General Nathanael Greene to ground, Cornwallis stripped down his heavy infantry to engage in an ultimately futile pursuit of the Continental Army into Virginia—the Race for the Dan River. Cornwallis subsequently wrecked what remained of his offensive potential in the campaign that culminated with the Guilford Courthouse

---

[12] John H. Beakes Jr., *Otho Holland Williams in the American Revolution* (Charleston, SC: Nautical and Aviation Publishing, 2015), loc. 52–182 of 9123, Kindle.

battle. After rehabilitating his command at Wilmington, North Carolina, Cornwallis ignored his stated mission to pacify the Carolinas in favor of his own pet scheme of invading Virginia to eliminate the Continental bases supporting the American insurgents. After some indecisive sparring with the smaller Franco-American army, Cornwallis fortified himself on the Williamsburg Neck to wait for reinforcements. Instead, the supporting Royal Navy squadron was driven off by a larger French flotilla, and Cornwallis's army endured a three-week siege before he surrendered on 19 October 1781. After his parole and return home, Cornwallis was publicly welcomed by King George III and, despite a bitter postwar exchange with Clinton, Cornwallis retained a large measure of public support. After a short stint as ambassador to the Prussian court, Cornwallis was made a knight companion and appointed as the governor-general and commander in chief of India in 1786. During his tenure, Cornwallis enacted reforms of the British East India Company as well as the civil service and justice systems. After successfully concluding the Third Anglo-Mysore War (1790–92), Cornwallis returned home to serve as the British master of the ordnance from 1794–98. Later, Cornwallis served as lord lieutenant of Ireland and represented the crown at the treaty ending the War of the Second Coalition (1798–1802). Reappointed governor-general of India, Cornwallis served only a short time before dying of a fever on 5 October 1805 at Gauspur, Ghazipur.[13]

Sergeant Roger Lamb

Although little is known of Roger Lamb's early life before his enlistment, his *Journal of the Occurrences during the Late American War*, which was published in 1809, shows evidence of an educated mind. Other entries in his journal reveal that Lamb had lived in Dublin, Ireland, before the war and indicate he had enlisted sometime in 1772. Lamb opened his journal on 3 April 1776, noting that he was a noncommissioned officer (NCO) with the *9th Regiment of Foot* on a ship sailing to Quebec City, Canada. While in Canada, Lamb saw his first action in a victorious skirmish at Three Rivers, which he described as "it really appeared to me to be a very serious matter, especially when the bullets came whistling by our ears . . . some of the veterans who had been well used to this kind of work said 'There is no danger if you hear the sound of the bullet . . . you are safe, and after the first charge all your fears will be done away'."[14] In June 1777, the *9th Foot* joined General John Burgoyne's army that was to march down the Hudson River valley to isolate New England from the middle states. Overburdened and unsupported, Burgoyne's army bogged down in fighting around Fort Ticonderoga, New York. Lamb saw much combat during this time and was briefly detailed as a field medic before rejoining the *9th Foot*, just in time for the humiliating surrender of Burgoyne's army at Saratoga on 17 October 1777. Burgoyne's disarmed troops were marched to a temporary internment camp in Cambridge, Massachusetts, while negotiations took place regarding their long-term status. When orders arrived in November 1778 to transfer the prisoners to Virginia, Lamb decided to escape—a risky proposition as he would have faced a court-martial as a deserter if caught and returned. Nevertheless, Lamb left the camp on the pretext of buying provisions with several other soldiers; after several days of trekking across the countryside, his party reached friendly lines in New York. After an extensive debriefing and an offer for an early return home, Lamb chose to reenlist in the *23d Regiment of Foot, the Royal Welch Fusiliers*. Not only was Lamb reappointed as a sergeant, he also was given a special bounty for his successful "honorable desertion" from American captivity. The *23d Foot* sailed as part of Lieutenant General Clinton's amphibious task force that captured Charleston in May 1780. The following August, Lamb's regiment took part in the major British victory at Camden. His first

---

[13] Charles Ross, ed., *Correspondence of Charles, First Marquis Cornwallis*, vol. 1 (London: John Murray, 1859), 1–17.
[14] Robert Lamb, *An Original and Authentic Journal of Occurrences during the Late American War from its Commencement to the Year 1783* (Dublin, Ireland: Wilkinson and Courtney, 1809), 104.

significant action was during the Race for the Dan, when he took part in a skirmish at McCowan's Ford on the Catawba River. At Guilford Courthouse, Lamb took part in the disjointed fight for the second line and personally helped save Cornwallis from riding into the enemy lines. After Guilford Courthouse, the *23d Foot* marched to Wilmington to refit, before taking part in Cornwallis's march into Virginia. After Cornwallis's capitulation at Yorktown, Lamb again escaped captivity and, after numerous misadventures, made his way to New York. After receiving his back pay, Lamb briefly clerked for the commandant of New York, a job "for which I had a good salary" before taking charge of a batch of new recruits for the *23d Foot*.[15] Lamb remained in New York until the *23d Foot* was repatriated, and returned to England in December 1783. Lamb was discharged in 1784 after having served 12 years in uniform; not enough time to qualify for the traditional half-pay on the retired rolls. Lamb returned to Dublin, became master of a Methodist free school, and in later years received a disability pension from Chelsea Hospital before dying in 1830.[16]

Major General Alexander Leslie

Born in 1731 to the Scottish Earl of Levin, Leslie was commissioned as a captain in the *64th Regiment of Foot* when it was organized in 1758. Leslie first saw action when he worked as an aide-de-camp during an expedition to Barbados in 1759. Captain Leslie rose in the ranks of the *64th Foot,* and by 1769 was the regimental lieutenant colonel. In February 1775, Leslie was ordered to march to Salem, Massachusetts, to confiscate a rebel cache of cannons and ordnance stores. Landing early one Sunday morning at the head of a battalion of 240 redcoats, Leslie found the townspeople alert and united in passively opposing his mission. A group of minutemen assembled and raised a drawbridge that blocked the only route to the cannons, giving a group of Patriot women time to move the cache to a different location. After a tense standoff with the local Patriot militia commander, Leslie was allowed to march his troops "fifty rods beyond the bridge and then return without molesting property" as a face-saving measure before withdrawing from the town.[17] Despite failing in his mission, Leslie was later given command of the *59th Foot*. After seeing service at Boston, Leslie joined Howe's army at New York, where, newly promoted as a brigadier general, he was described as a "genteel little man, lives well and drinks good claret."[18] Leslie was given command of an infantry brigade, which he successfully led during the Battles of Long Island and Princeton. After his promotion to major general in February 1779, Leslie commanded a diversionary attack in the Chesapeake Bay during Clinton's campaign to take Charleston. Leslie sailed south to Charleston in late 1780 with reinforcements for Cornwallis, which joined up with Cornwallis's army after Cowpens. In Cornwallis's army, Leslie was given command of a brigade comprised of the *2d Battalion, 71st Regiment of Foot,* and the *Von Bose Regiment.* During the Guilford Courthouse battle, Leslie seems to have exercised little control of his brigade after the opening engagement, as his regiments broke away from the main British line in pursuit of Lee's Corps of Observation. After the Battle of Guilford Courthouse, Leslie returned to Charleston to serve as its garrison commander and ended the war as the overall commander in the South after Cornwallis's march into Virginia. Leslie oversaw the evacuation of Charleston in December 1782 before returning home to Britain due to poor health. After the war, Leslie served as second in command to royal forces in Scotland, where he was seriously injured by a mob in the aftermath of a mutiny. After he was "seized with a dangerous illness," Leslie passed away on 27 December 1794.[19]

[15] Lamb, *An Original and Authentic Journal of Occurrences During the Late American War from its Commencement to the Year 1783*, 414.

[16] Lawrence E. Babits and Joshua Howard, *Long, Obstinate and Bloody: The Battle of Guilford Courthouse* (Chapel Hill: University of North Carolina Press, 2009), 200–1.

[17] Charles M. Endicott, *Account of Leslie's Retreat at the North Bridge on Sunday February 25th 1775* (Salem, MA: William Ives and George W. Pease Printers, 1856), 26–29.

[18] Babits and Howard, *Long, Obstinate and Bloody*, 86.

[19] Babits and Howard, *Long, Obstinate and Bloody*, 196–97.

Brigadier General Charles O'Hara

Born in 1740 as the illegitimate son of the British envoy to Portugal, O'Hara attended Westminster School in his youth. In 1752, his father purchased a cornet's commission in the *3d Dragoons* for him, and in 1758, a lieutenant's commission in the *2d Regiment of Foot*, the *Coldstream Guards*. While in the *Coldstream Guards*, he saw limited action in the Seven Years' War (1756–63) before running afoul of a senior officer, which resulted in his assignment as the governor of Senegambia.[20] With that post, O'Hara assumed command of the *African Corps of Foot*, essentially a unit made of criminal soldiers sentenced to life service in Africa in lieu of a death sentence. In 1776, O'Hara was recalled to Britain after the Board of Trade tired of his mismanagement. O'Hara deployed to America in March 1777, where he served as a member of General Sir William Howe's staff in New York. O'Hara returned to Britain in February 1779, where he languished until selected to command the *Brigade of Guards*, which at that time was garrisoning New York. Although O'Hara was given his new assignment in May 1780, he did not assume command in North American until December 1780, when he joined the *Brigade of Guards* in Charleston. From Charleston, O'Hara's brigade marched inland to reinforce Cornwallis's army, linking up shortly after the Cowpens battle. During the Race for the Dan, O'Hara's infantry formed the vanguard of Cornwallis's army. When Cornwallis attempted a surprise crossing at Cowan's Ford, O'Hara braved the current and American rifle fire to lead his troops across the river. At the Guilford Courthouse battle, O'Hara exercised direct command over the *1st Battalion, Brigade of Guards*, and separate *Guards* companies, as the *2d Battalion, Brigade of Guards*, was detached to reinforce the British right wing. O'Hara was in the thick of the fight at the second line, receiving a musket ball wound while rallying the *2d Guards* battalion. After a quick stop for a field dressing, O'Hara led the *2d Guards* battalion and separate companies into attacking the American third line. When the *Guards* were repulsed by the 1st Maryland Regiment, O'Hara was shot in the chest—and may have been captured for a brief period before being retaken by a *Guards* counterattack. O'Hara survived his wounds and resumed command of the *Guards* during the rest of Cornwallis's failed campaign in Virginia. At the surrender ceremony at Yorktown, O'Hara had the dubious honor of formally surrendering the British Army when Cornwallis pleaded sickness to avoid surrendering in person. After his parole, O'Hara returned to New York, where he was promoted to major general before sailing to a new command in the Caribbean. After returning to Britain in late 1782, O'Hara took a leave of absence to hide out in France until 1785 to avoid gambling debts, which were generously paid off by Cornwallis. Thus restored, O'Hara reentered active service as the commander, and later lieutenant governor, of the Gibraltar garrison. In 1793, O'Hara was taken prisoner when French forces led by Napoleon Bonaparte captured the island of Toulon. After two years of injury and deprivation while in prison, O'Hara returned to Gibraltar as governor after a prisoner exchange. O'Hara was promoted to full general in 1798 and remained in Gibraltar until his death on 25 February 1802.[21]

Lieutenant Colonel Banastre Tarleton

Born in 1754 to a middle-class Liverpool family, Banastre Tarleton enrolled at the University College, where he spent more time engaged in athletics and gambling than studying law. After running out of money, he obtained a cornet's commission in the *1st Regiment of Dragoon Guards*, using money borrowed from his mother. In 1775, Tarleton volunteered for service in America and quickly earned a reputation as a daring battlefield commander. His performance during the campaigns of 1777 and 1778, particularly his capture of American General Charles Henry Lee in a bold nighttime

---

[20] Used from about 1765, the term *Senegambia* refers to British settlements on Saint-Louis and the Island of Gorée in Senegal and James Island in the Gambia. "Senegambia Origins," Gambia Information Site, accessed 1 April 2020.
[21] Babits and Howard, *Long, Obstinate, and Bloody*, 195–96.

raid, earned promotion to lieutenant colonel. Tarleton was also given command of the *British Legion*, a combined arms regiment of infantry and dragoons raised under the Provincial system. In 1780, Tarleton's *Legion* was deployed as part of Clinton's expedition to South Carolina, although his troops were temporarily unhorsed when the transport carrying their mounts foundered in an Atlantic gale. Once in the South, Tarleton's troops had to march overland from Savannah to Charleston, confiscating mounts along the way. On 12 April 1780, Tarleton led a force of provincials and regulars in a surprise nighttime attack that captured the key Continental position at Monck's Corner, which guarded the main road into Charleston. With that audacious win, Tarleton cemented a reputation as a bold and exceedingly dangerous battlefield commander. At the Battle of Waxhaws, Tarleton's *Legion* destroyed a larger Virginia infantry regiment, giving rise to stories that the Tories ignored a white flag and killed defenseless Continentals. Regardless of the truth, Tarleton was forever afterward branded as a brutal butcher of helpless prisoners. During the pacification phase of Cornwallis's operation in South Carolina, Tarleton repeatedly bested Patriot guerrilla bands under Thomas Sumter's command, although Tarleton was never able to defeat Francis Marion's much better disciplined partisan force. On 17 January 1781, Tarleton's light infantry was largely wiped out at the Cowpens by Daniel Morgan's light infantry corps, although Tarleton succeeded in escaping with most of his mounted troops. Afterward, Tarleton received much criticism by other British officers, particularly his subordinate commanders, who believed the defeat at the Cowpens was due to his lack of maturity and experience. Surprisingly, Cornwallis did not blame Tarleton for the disaster, and he was to remain in command of the much smaller *Legion* during the remainder of the Guilford Courthouse campaign, where his hand was maimed by a rifle ball in the closing stages of the battle. During Cornwallis's march through Virginia, Tarleton came within minutes of capturing Governor Thomas Jefferson and the legislative body of Virginia, but instead had to settle for destroying abandoned ordnance stores. After surrendering his command at Yorktown, Tarleton suffered humiliation when he was pointedly excluded from an invitation for Cornwallis's officers to dine with the American and French officers. Paroled to England, Tarleton indulged in excesses such as gambling, drinking, and a series of mistresses that landed him deeply in debt. Politically ostracized by the Tory-dominated Parliament over a new job, Tarleton caused further problems when he published his *A History of the Campaigns of 1780 and 1781 in the Southern Provinces of North America* (1787), in which he minimized his own errors and pinned much blame for the failed southern campaign on Cornwallis's poor generalship. Tarleton's book generated much bad publicity and prompted Roderick Mackenzie, a former lieutenant of the *71st Regiment of Foot* who was captured at Cowpens, to publish a detailed rebuttal, *Strictures on Lt. Col. Tarleton's History* (1787). Tarleton served in Parliament from 1790 to 1812, during which he retained his Army commission but saw no major service during the Peninsular War (1808–14) or War of 1812. Despite his lack of later war service, Tarleton was promoted to major general before dying in January 1833.[22]

Lieutenant Colonel James Webster

James Webster was born in 1743 to a prominent minister's family in Edinburgh, Scotland. Webster was commissioned a lieutenant in the *33d Regiment of Foot* in 1760, and by 1776 he was the lieutenant colonel in command when deployed to America. The regimental colonel was Cornwallis, and the two men developed a close professional relationship, with Webster as a trusted subordinate. Webster performed with distinction at Monmouth in 1778 and was given command of an ad hoc brigade, with the *23d* and *33d Foot*, for the Southern Campaign. Webster continued to excel in the South, support-

<hr>

[22] "Lieutenant Colonel Banastre Tarleton," Cowpens National Battlefield, South Carolina, National Park Service (website), accessed 8 April 2020.

ing Tarleton's attack at Monck's Corner in April 1780, and forming the main effort of Cornwallis's attack at Camden in August 1780. During the Race for the Dan, Webster skillfully adapted his troops to their new role as light infantry. At the Whitesell's Mill skirmish of 6 March 1780, Webster came within minutes of cutting off the withdrawal of the American light infantry force, even riding his horse into Reedy Fork Creek while under American rifle fire as an example to his troops. At Guilford Courthouse, Webster led his brigade from the front, continually risking enemy fire to motivate his troops to break both Eaton's North Carolina brigade and Lawson's Virginia Line brigade. Webster's luck ran out when he led the *33d Foot* in an unsupported attack on the American third line and was seriously wounded by a Continental musket ball to the knee. Webster lived after the battle for two weeks, dying probably from gangrene or secondary infection, and he was interred near Elizabethtown, North Carolina. Sergeant Lamb recorded Cornwallis's reaction at the news of Webster's death: "It was reported in the army that when Cornwallis received the news of Webster's death, his lordship was struck with a pungent sorrow, then turning himself, he looked upon his sword, and emphatically exclaimed, 'I have lost my scabbard'."[23]

---

[23] Babits and Howard, *Long, Obstinate, and Bloody*, 82–85, 174.

APPENDIX B

FIGURE 5–ORDERS OF BATTLE

British Army
~ 2,200 with 1,900 engaged at the battle

LtGen Lord Charles Cornwallis

The Brigade of Guards
BGen Charles O'Hara (481)

Leslie's Brigade
MajGen Alexander Leslie
(565)

Webster's Brigade
LtCol James Webster (472)

1st Battalion Guards
LtCol Chapple Norton (180)

2d Battalion
71st Highlanders
Capt R. Hutchison (244)

23d Foot
Capt Thomas Peter (238)

2d Battalion Guards
LtCol James Stuart (180)

Von Bose Regiment
Maj Johann du Buy (244)

33d Foot
Capt Frederick Cornwallis (238)

2d Battalion Guards
LtCol James Stuart (180)

Guards Light Infantry Company (90)

Guards Grenadier Company (90)

British Legion
LtCol Banastre Tarleton (272)

Ansbach Jäger Company
Capt Friedrich von Roeder (84)

Royal NC Regiment of Loyalists
LtCol John Hamilton
(140)

Royal Artillery
4x 6–pounders
2x 3–pounders
(50)

Combat Trains

Source: Babits and Howard, *Long, Obstinate, and Bloody: The Battle of Guilford Courthouse,* 219–21.

American Army
~ 4,400 with 4,000 engaged at the battle

MajGen Nathanael Greene

Maryland Continental Brigade LtCol Otho Williams (400)	Maryland Continental Brigade BGen Isaac Huger (900)	Corps of Observation LtCol Henry Lee	Corps of Observation LtCol Washington
1st Maryland Continental Regiment Col John Gunby (400)	1st Virginia Continental Regiment Col Samuel Hawes (450)	Partisan Legion (240)	Provisional Legionary Corps (238)
2d Maryland Continental Regiment LtCol Benjamin Ford (400)	2d Virginia Continental Regiment Col John Green (450)	Rifle Battalion Col William Campbell (100)	Rifle Battalion Col Charles Lynch (150)
Virginia Line Brigade BGen Edward Stevens (750)	Virginia Line Brigade BGen Robert Lawson (650)	Virginia Light Infantry Company (60)	Continental Light Infantry Companies (120)
Col Peter Perkins (200)	Col Beverley Randolph (200)	North Carolina Militia BGen John Butler (600) –4 Battalions –3 Companies	North Carolina Militia BGen Thomas Eaton (600) –3 Battalions
Col Nathan Cocke (200)	Col John Holcombe (250)		
		Continental Artillery 4x 6–pounders (40)	Combat Trains
Maj John Pope (150)	Maj Henry Skipwith (150)	North Carolina Continental Infantry Company (40)	
Maj Alex Stuart (200)			

Source: Babits and Howard, *Long, Obstinate, and Bloody: The Battle of Guilford Courthouse*, 219–21.

APPENDIX C

CHRONOLOGY OF MAJOR EVENTS

Date	Event
February–May 1780	Siege of Charleston
12 May	Charleston surrenders (British victory)
29 May	Battle of Waxhaws and Buford's Massacre (British victory)
30 May	Surrender of Thicketty Fort, SC, to Patriot force
16 August	Battle of Camden, SC (British victory)
18 August	Battle of Musgrove's Mill, SC (Patriot victory)
18 August	Battle of Fishing Creek/Catawba Ford (Loyalist victory)
25 September	Cornwallis occupies Charlotte, NC
7 October	Battle of Kings Mountain, NC (Patriot victory)
17 January 1781	Battle of Cowpens, SC (American victory)
18 January	Leslie joins Cornwallis with reinforcements
19 January	Cornwallis begins pursuit of Morgan
24 January	Cornwallis burns his baggage trains at Ramsour's Mill, NC
30 January	Morgan's flying army consolidates with Greene's main body
1 February	Skirmish at Cowan's Ford, NC
2–3 February	Greene evacuates factory at Salisbury, crosses the Yadkin River, NC
4 February	Greene's army camps at Guilford Courthouse, NC
14–15 February	Greene's army wins the Race for the Dan
16 February	Cornwallis withdraws to Hillsborough, NC
18 February	Williams's light infantry corps recrosses the Dan
22 February	Greene redeploys his main body back into North Carolina
25 February	Lee's Partisan Legion defeats Pyle's Regiment (Pyle's Hacking Match)
26 February	Cornwallis relocates his command to Stinking Quarter Creek, NC
6 March	Skirmish at Whitesell's Mill, departure of Colonel Pickens/SC militia
10 March	Greene receives North Carolina and Virginia militia reinforcements
14 March	AM—Greene camps his command at Guilford Courthouse
15 March	PM—Cornwallis locates Greene, plans attack the next morning
(times are approximate)	0300—Cornwallis begins movement to Guilford Courthouse
	0400–0500—Dragoon skirmishes at New Garden Meeting House
	0600—Greene feeds and deploys his men into three battle lines
	1000—Lee's Corps of Observation withdraws into position
	1200—Cornwallis's army arrives at Guilford Courthouse
	1300—British advance on first line
	1315—First line breaks
	1345—Fight at the second line
	1445—Webster begins attack on the third line
	1500—Fight between *Guards* and 1st Maryland
	1530—Greene's main body breaks contact and withdraws

SELECTED GLOSSARY

The following glossary is not comprehensive, but it is intended to briefly define military-specific or archaic terms used in the body of the text.

Abatis: a field obstacle made of sharpened tree branches oriented toward the enemy.

Area defense: a defensive task by which a unit focuses on denying enemy forces access to designated terrain for a specific period of time instead of trying to destroy the enemy.

Artificer: a skilled craftsman employed to make or repair military equipment and weapons.

Bateau/bateaux: a small flat-bottomed wood boat, double-ended and of shallow draft construction that made it useful for carrying heavy cargo and supplies.

Bayonet: a long stabbing blade affixed, usually by a metal socket, to the muzzle of a military musket.

Case shot: antipersonnel munition fired from a muzzle-loaded cannon. Case shot consisted of small golf ball-size projectiles loaded in an iron or tin canister, which, when fired, would fragment at the muzzle to produce a shotgun-like blast.

Chevaux-de-frise: a series of anticavalry obstacles, made of wooden logs with projecting wood spikes.

Commissary: a military officer in charge of the procurement, delivery, and distribution of provisions, rations for soldiers, and fodder and forage for animals.

Delay: a type of defense where a unit under pressure trades terrain for time, slowing down the enemy force and inflicting maximum damage while avoiding a decisive engagement.

Doctrine: fundamental principles, including tactics, techniques, and procedures (TTP), and a common language by which a military force guides its actions in support of national objectives.

Dragoon: a mounted fighting soldier. Unlike cavalry, which were trained to fight from horseback, dragoons were originally intended as mounted infantry, riding to maneuver then dismounting to fight. By the revolution, dragoons were employed as conventional cavalry, generally fighting from horseback with the saber as primary weapon.

Fascines: a bundle of sticks or branches tied together, used to fill in swampy ground, or to reinforce the side of an entrenchment.

Flintlock: a firearm ignited by a spark from a lock mounted flint.

Flying army: a term first used by General Greene in a letter to George Washington to describe General Daniel Morgan's corps of light infantry and dragoons. The flying army was expected to draw provisions from the countryside, and thereby unburdened by supply wagons, the unit was capable of rapidly outrunning all enemy formations with the exception of Lieutenant Colonel Banastre Tarleton's *British Legion*.

Forage: 1) bulk food for horses and cattle, usually grass or hay; 2) the act of searching for food.

Grapeshot: an antipersonnel munition, consisting of large iron balls contained in a canvas bag held together by rope, metal bands, or wood.

Guard: a security task in which a unit protects the main force by fighting to gain time and reporting information about the enemy, while preventing the enemy from directly observing or engaging the main body.

Hessian: German auxiliary units contracted for military service by the British. Commonly named for the German states of Hesse-Kassel and Hesse-Hanau but recruited from other German states.

Jäger: literally, a hunter; jäger were German light infantry, recruited from gamekeepers and hunters who worked for noble landholders. The jäger were armed with light muzzle-loading rifles and trained to fight as individual skirmishers.

Legion: a mixed-arm regiment, usually of dragoons and infantry, possibly including artillery.

Logistics: the art and science of moving, feeding, clothing, housing, and resupplying military units.

Loyalist: American colonists loyal to the British monarchy during the Revolution; sometimes called Loyal Americans.

Lunette: a minor fortification with two faces forming a projecting angle.

Magazine: a building or structure designed to segregate gunpowder and cartridges in isolation from other military stores, lessening collateral damage from an explosion. During the American Revolution, _magazine_ denoted an intermediate logistics node, a location where bulk commodities were shipped in by river and distributed to the tactical units by wagon or cart.

Militia: a military force raised from a civilian population for short-term service.

Musket: a muzzle-loaded smoothbore firearm. During the revolution, muskets were ignited by a flintlock firing system.

Operation: the coordinated military actions of a state, operating according to a plan, to meet a specific strategic goal.

Ordnance: military-specific equipment and materiel, to include weapons, cannons, and munitions.

Patriot: to Britons, the term _patriot_ was disparaging, describing a deluded, rebellious fool. In America, the label _Patriot_ was adopted as a mark of honor for a person advocating American independence from British rule.

Provincials: a practice dating back to the French and Indian Wars (1754–63), British governors and department commanders organized short-service provincial Loyalist regiments to augment regular British units. Officers and soldiers were mustered only for the duration of the conflict and were not entitled to pensions or other benefits enjoyed by establishment units.

Pursuit: an offensive military task designed to overtake or cut off an enemy force attempting to escape, with the intent of destroying the enemy.

Quartermaster: a military officer, or appointed civilian, responsible for procurement and distribution of supplies such as clothing, coordination of transportation, and construction of troop housing in camps.

Reconnaissance: a mission to obtain information about the activities and state of an enemy force. Done by visual means, interrogation of civilians or enemy prisoners, and analysis of information.

Redan: a V-shaped salient angle pointed toward the expected enemy approach. Used to allow defenders to enfilade an enemy attacker with fire.

Regiment: the highest permanent tactical unit employed during the revolution. Led by a colonel, assisted by a lieutenant colonel and major, a regiment could have between 5 and 10 subordinate companies. Regiments were normally of a single arm, usually infantry or dragoons.

Saber: a heavy sword with a curved blade and single cutting edge, usually used by mounted troops.

Strategy: planning, coordination, and general direction of military operations to meet national objectives.

Tactics: the employment and deliberate arrangement of combat forces in relation to each other to accomplish a military objective.

Tory/Tories: Patriot Americans co-opted the political term _Tory_ from its original usage to describe the political party that supported royal policies. The Irish roots of _Tory_ denoted an outlaw or criminal; so, the term was adopted as a derogatory label for supporter of the king.

Vanguard, or Van: a detachment of light infantry and dragoons marching in the advance of a large body of infantry. The vanguard performed intelligence gathering and screened the main body from surprise attack.

Withdrawal: a force in combat breaks contact and moves out of range of the enemy force.

Whig/Whigs: American revolutionaries were sometimes described as Whigs, a disparaging term (country bumpkin) used from the mid-seventeenth century to describe the English political opponents of the king.

ABOUT THE AUTHOR

Harold Allen Skinner Jr. currently serves as the command historian for the U.S. Army Soldier Support Institute (SSI), located at Fort Jackson, South Carolina. From August 2019 to January 2020, Skinner deployed under the Civilian Expeditionary Workforce program as the theater historian for Combined Joint Task Force–Operation Inherent Resolve (CJTF-OIR). He was awarded the Joint Civilian Service Commendation Award for his theater historian accomplishments. From April 2015 to July 2019, Skinner was the command historian for the 81st Division (Readiness), U.S. Army Reserve, at Fort Jackson. He retired as an Army major from the Active Guard/Reserve program in 2015 after 25 years of service with six operational tours, which included deployments in support of Operations Desert Storm and Iraqi Freedom. Previous military assignments included military history detachment commander, task force historian, and command historian for the Indiana National Guard and the 38th Infantry Division. Skinner received a bachelor of science in administration of justice from the Southern Illinois University in Carbondale and a master of military art and science in military history from the U.S. Army Command and General Staff College in Fort Leavenworth, Kansas. He is currently enrolled in the history PhD program at Liberty University in Lynchburg, Virginia. Skinner's most recent published work, *The Staff Ride Handbook for the Battle of Kings Mountain —7 October 1780*, was released by the Army University Press in April 2020 and was later selected as a finalist for a 2020 Army Historical Foundation Distinguished Writing Award. Other recent works include a series of peer reviewed articles for *1914–1918 Online: The International Encyclopedia of the First World War*, and an article published in the *Army Chaplain's Journal* telling the story of the first Army chaplain awarded the Medal of Honor.